Instrumentation for Engineers

J. D. Turner

Instrumentation
for Engineers

Springer-Verlag New York Inc.

© J. D. Turner 1988

All rights reserved. No reproduction, copy or transmission of this publication may be made without written permission.

First published 1988

Published by
MACMILLAN EDUCATION LTD
Houndmills, Basingstoke, Hampshire RG21 2XS
and London
Companies and representatives throughout the world

Sole distributors in the USA and Canada
Springer-Verlag New York Inc.
175 Fifth Avenue,
New York, NY 10010
USA

Printed in China

Library of Congress Cataloging-in-Publication Data
Turner, J.D.
 Instrumentation for engineers,

 Bibliography: p.
 Includes index.
 1. Engineering instruments. I. Title.
TA165.T87 1988   620'.0044   88–18476
ISBN 0–387–91333–5

For Caroline and Jack

Contents

Preface   xi

Acknowledgements   xiii

1 The Performance of Instrumentation Systems   1
    Introduction   1
    Generalised instrumentation design   3
    The performance of instrumentation systems   6
    Error analysis   6

2 Sensors and Transducers   13
    Introduction   13
    Displacement sensing   13
    Velocity sensing   21
    Acceleration sensing   24
    Strain measurement   25
    Flow sensors   29
    Temperature sensors   32
    Optical sensors   35
    Acoustic sensors   38
    Hall effect sensors   39

3 Signal Conditioning   41
    Introduction   41
    Bridge circuits   42
    Operational amplifier signal conditioning circuits   55
    Analysing op-amp circuits   62

4 Analogue Filters — 79
Introduction — 79
Filter order — 81
Filter class — 82
Operational-amplifier filters — 84
Special-purpose filter devices — 95

5 Signal Conversion — 102
Introduction — 102
Digital and analogue conversion fundamentals — 103
Digital-to-analogue converters — 105
Frequency-to-voltage converters — 109
Sample-and-hold devices — 109
Analogue-to-digital converters — 110
Analogue multiplexers — 115
Example design — 116

6 Digital Circuits and Microprocessor Interfacing — 118
Introduction — 118
Digital device families — 118
Combinational logic, gates and Boolean algebra — 121
Sequential logic circuits — 125
Digital systems interfacing — 130
Number codes — 135
Microprocessors — 138
Example interface designs — 144
Communication standards — 157

7 Frequency Domain Analysis — 159
Introduction — 159
The modal domain — 164
Waterfall diagrams — 167
Vector response diagrams — 168
Fourier analysis — 170
Fourier series — 170
The Fourier Transform — 177

8 Practical Spectrum Analysis — 180
Introduction — 180
Analogue analysers — 180
Digital analysers — 182
The Fast Fourier Transform — 184
Aliasing and Shannon's Sampling Theorem — 188

Windowing	190
Choice of window	195
Glossary of FFT analyser terminology	199

9 Correlation and Spectral Analysis — **203**
- Introduction — 203
- Signal classification — 203
- Autocorrelation — 205
- Interpreting autocorrelation diagrams — 208
- Cross-correlation — 213
- Interpreting cross-correlation functions — 215

Bibliography and Further Reading — 217

Index — 219

Preface

The science (or even the art!) of instrumentation is of fundamental importance to engineers, scientists and medical workers. Instruments are the eyes and ears of the technologist. (His nose is reserved for detecting the effects of excess current.) Without sensors and their associated signal processing systems there would be no modern transport, no National Grid distributing electricity, and anyone unlucky enough to fall ill would be offered only the most primitive medical treatment. The progress that has been made in almost all areas of technology can be seen in terms of the rate at which the necessary instrumentation has been developed. For example, in recent years many improvements have been made to the performance of the internal combustion engine. More and more power has been squeezed out of smaller and more economic engines. One of the reasons is that in the last few years sensors have been developed which allow investigations to be made of the way in which the flame front spreads inside a cylinder after ignition. This work has led to a redesign of the geometry of the inlet valves and the piston, and more efficient engines are the result.

The process of instrumentation is often considered solely in terms of the sensors used and their associated electronics. However, there are two steps involved in making any measurement. These are, first, getting the data, which is where sensors and electronics are used, and second, analysing it. The analysis may simply consist of an assessment of the errors, or may involve the application of some more complicated technique such as Fourier or correlation analysis. The important point is that some form of signal analysis is almost always necessary. The type of analysis to be applied may well have implications for the kind of sensors used, or vice versa. Thus, an instrumentation system should be designed or used with due regard to the expectations that the user has of the results.

In view of the importance of instrumentation, it is vital that engineers and scientists have a good working knowledge of the principles involved in measurement and analysis. If the best possible performance is to be had from a measurement system then its weaknesses must be thoroughly understood, as well as its strong points. It is often said of electronic systems and devices that the most important features to look out for are the ones that the manufacturer has NOT included on the data sheet!

The basis of this book is a course in instrumentation which I give to final year mechanical engineering undergraduates. This course presents instrumentation from the viewpoints of both electronics and signal analysis. The sensors and electronic circuits likely to be needed by a final year student project and for postgraduate research are comprehensively covered. Three chapters are devoted to methods of time and frequency domain analysis, and suggestions for further reading on some of the more specialised techniques are given in the bibliography. It is therefore a suitable university text for students of engineering, science or medicine who are seeking a practical guide to instrumentation. It is also hoped that the book will be of use to practising engineers in general.

Finally, some words of thanks. I would like to record my gratitude to my friend and colleague Tony Pretlove, who first showed me the fascination of engineering, and turned a reluctant physicist into a happy engineer. Thanks are also due to Martyn Ramsden, who read the manuscript and offered many helpful suggestions, and to Professor Roy Farrar of Southampton University who encouraged the work. The text was typed and corrected by myself on that wonderful thing, a word-processor program on a microcomputer. Without the aid of this software it would have been impossible to keep the end result in View.

Hear ye not the hum of mighty workings?
                              Keats: to Haydon

Southampton University   J. D. TURNER

Acknowledgements

The author wishes to thank Professor R. Farrar, Head of the Department of Mechanical Engineering at Southampton University, for his encouragement regarding the lecture courses on which this book is based, and Mr A. Munday and Dr G. Pitts, also of Southampton University, for their many helpful suggestions about the content of the book.

Permission for the reproduction of figures and tables is acknowledged from the following bodies:

The Parker Publishing Company Inc for tables 4.1, 4.2, 4.3, 4.4, 4.5, 4.6, 4.7 and 4.8. E. G. and G. Reticon for the datasheets for the RF5609, RM5604, RM5605 and RM5606, which appear at the end of chapter 4. Texas Instruments Ltd for the datasheet on the 741 op-amp at the end of chapter 3. Brüel and Kjaer (UK) Ltd for figures 2.11, 2.13 and 8.21. Butterworth & Co (Publishers) Ltd for figures 6.23 and 6.24.

Every effort has been made to trace all copyright holders but if any have been inadvertently overlooked the publishers will be pleased to make the necessary arrangement at the first opportunity.

Chapter 1
The Performance of Instrumentation Systems

> Only a simpleton believes what he is told.
> A prudent man checks to see where he is going.
> (Proverbs 14:15)

INTRODUCTION

Instrumentation is a subject which is of fundamental importance to engineering, and to almost all of the practically-based sciences. From the student undertaking a laboratory investigation to the operators of a nuclear power plant, accurate measurements are an essential prerequisite to the understanding and control of all physical processes.

In general, an instrumentation system may fall into one of two categories. First, there are the laboratory or experimental measurement techniques used for research and development. This classification includes the instruments used to study the performance of an engineering prototype, and the laboratory devices used where high precision is required. The most important consideration faced by the designer of a measurement system intended for experimental work is performance. In acquiring research data a high degree of repeatability, accuracy, linearity and reliability are required. The cost of this kind of system is usually of secondary importance.

The second sort of measurement system is that which forms part of a well-understood device, usually a commercial product. Examples of this kind of instrument can be seen in any motor vehicle. The driver is provided with a speedometer to help control the vehicle, a petrol gauge to indicate when fuel is required, and a mileometer or other indicator to show when maintenance is needed.

For a well-understood system such as a motor vehicle, a lower degree of instrument performance than that required for research is usually sufficient. For example, the accuracy of the average automotive speedometer is probably no more than ±10 per cent. However, this lack of resolution is entirely adequate to control the vehicle. In general, the instrumentation supplied as part of a mass-produced device is of poorer quality than that used for experimental work. The principal reason for this is so that the complete system can be produced at an economic cost.

In the automotive example above, three kinds of information were identified:

(a) parameters used for control, such as speed;
(b) warning devices, such as a fuel gauge;
(c) condition monitoring data, such as that provided by a mileometer or hours-run indicator.

Some at least of these categories of information are common to all measurement devices forming part of a well-understood system.

Measurements are carried out by means of transducers. A transducer is a device which converts energy from one form to another. Output transducers or actuators convert electrical, pneumatic or hydraulic energy into mechanical force. Input transducers or sensors convert state parameters such as temperature, pressure, force, magnetic field strength etc. into (usually) electrical energy, since this is generally the most convenient form for measurement or signal processing. Since the subject of this book is instrumentation we shall be mainly concerned with sensors rather than actuators.

Once information regarding the variation of a physical parameter has been produced in electrical form by the action of a sensor, the instrumentation designer has then to consider a number of subsidiary problems. The output of most sensors takes the form of a small electrical signal. This signal is most often a varying voltage, but can also be a varying current or charge. The voltage form is usually the most convenient for subsequent analysis. Thus, appropriate signal conditioning circuits have to be constructed to convert the signal to voltage form if necessary, amplify it, and apply filtering to remove unwanted noise.

At this stage information about the behaviour of the parameter of interest is available to the engineer in the form of an analogue voltage. A decision must now be made regarding the best way to proceed. In the past, much signal analysis was carried out by means of analogue circuits. For example, until the mid-1970s it was common to examine the spectrum of a signal by means of analogue devices such as the swept-filter analyser. Although analogue techniques are still sometimes used, the availability of digital equipment at relatively low cost has meant that signals are now commonly converted from analogue to digital form if anything more than rudimentary signal processing is required. For example, almost all commercial spectrum analysers work by converting a sampled analogue input to digital form. An integration algorithm is then used to carry out a Fourier Transform on the data.

The instrumentation designer has therefore to decide whether to retain his data in analogue form, or whether the required signal processing and analysis will be better carried out in a digital format. There is no doubt that the digital approach greatly facilitates signal analysis, but the process of digitisation is not always straightforward. Care has to be taken over aliasing, sample rate and the choice of binary word size (which determines resolution).

The kind of signal processing ultimately carried out will depend upon the information required from the signal. Frequently all that is needed is the amplitude of the signal, to indicate the amplitude of the force, pressure etc. being measured. In such cases a straightforward display of the signal on a moving coil meter or oscilloscope is often all that is necessary. However, it is often necessary to use signal processing techniques such as smoothing, averaging, correlation or the Fast Fourier Transform mentioned earlier to extract the required information.

GENERALISED INSTRUMENTATION DESIGN

An instrument can be defined as a system which maintains a prescribed relationship between the parameter being measured and some other physical variable. The second variable is used as a means of communicating information regarding the first, either to a human observer or to some other measurement or control system. An indication of how well a prescribed functional relationship is maintained can be had from the static and dynamic calibration of the instrument concerned.

A measurement system may be broken down into its functional elements as shown in figure 1.1. Every instrumentation system contains some or all of these functional blocks. If the behaviour of the elements is known, an assessment of the performance of the complete system can be made.

Information regarding the state of a physical system is obtained through a change in one of the properties of the system. For example, in a vibrating system changes in the amplitude, frequency or phase of vibration all convey information about the state of the system. The physical parameter being measured, which forms the input to the generalised measurement system of figure 1.1, is known as the measurand. The primary sensing elements in any measuring system are those which first receive energy from the object being measured, and which produce an output according to a well-understood relationship with the measurand.

The measurand is always disturbed by the act of measurement, although well-designed instruments are arranged so that this effect is minimised. The disturbing effect of the act of measurement can be illustrated by the following examples. Consider the vibrating system discussed above. The usual way of instrumenting such a system is by means of accelerometers. However, accelerometers have a mass of 100 grammes or so, and the effect of adding mass to a vibrating system is to alter its dynamic characteristics. The extent

4 INSTRUMENTATION FOR ENGINEERS

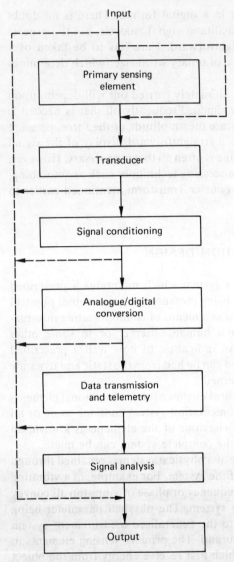

Figure 1.1 The components of an instrumentation system.

of the alteration depends on the relative masses of the vibrating system and the sensors.

At the other end of the scale, the atomic structure of a crystalline material such as silicon may be investigated by means of an electron microscope. Unfortunately, the high-energy electrons used to form an image of the material can displace atoms from the crystal being studied. (At the level of quantum effects, this is a consequence of Heisenberg's Uncertainty

Principle). Once again, we see that the act of measurement always changes the system under investigation to some extent.

The energy output from the primary sensing element is passed to a transducer, or device which converts energy from one form to another. The output of almost all practical transducers is electrical, since signals in an electrical form can most conveniently be processed or transmitted. There are three fundamental types of transducer, as shown by figure 1.2.

In the first type the same form of energy exists at both the input and the output. This kind of transducer is known as a modifier, since energy is modified rather than converted to a different form. An example might be a bandpass filter, which can be used to measure the energy within a particular bandwidth in an electrical signal.

The second category of transducers are those known as self-generating or self-exciting. In these devices electrical signals are produced directly from a non-electrical input, without the application of any external energy. Examples of self-generating sensors are thermocouples, photovoltaic cells, and devices such as accelerometers which depend for their operation on the piezoelectric effect. Self-generating transducers are usually characterised by their very low output energy. They are normally followed by a number of amplification stages to increase the signal amplitude to a useful level.

The third category of transducers also produces an electrical output from a non-electrical input, but requires an external source of energy to

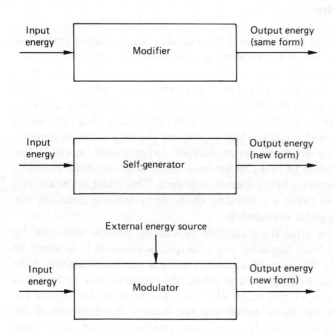

Figure 1.2 Classification of transducers into three types.

function. These are known as modulating transducers. The strain gauge is a familiar example, in which a mechanical deformation is used to control a variable resistance and a signal voltage is generated by passing a current through the device.

THE PERFORMANCE OF INSTRUMENTATION SYSTEMS

The performance of any instrument depends upon both its static and its dynamic characteristics. In the case of rapidly varying quantities, the relationship between the input and the output of an instrument is usually expressed by means of differential equations. However, for many applications the parameters being measured vary sufficiently slowly for the dynamic effects to be neglected.

The errors caused by non-linearity, drift, resolution errors and repeatability are usually considered as static characteristics. Hysteresis, settling time and variations in the response at different frequencies are dynamic effects. The total system response is obtained by combining the static and dynamic responses.

ERROR ANALYSIS

All instruments suffer from inherent inaccuracies. Every measurement has an associated error (except the counting of numbers, which obviously has no error). To estimate the uncertainty in a measurement it is necessary to know the form of the error. In general, any error is a combination of two kinds: a systematic error, in which all the readings are systematically shifted in one direction, and a random error in which repeated readings are found to be scattered around a mean which represents the true value. Systematic errors usually arise from an unsatisfactory experimental method. For example, it is common to find that an instrument does not quite return to zero when the parameter being measured is zero. This effect is known as a zero error, and will cause a systematic shift—or systematic error—in any reading made using that instrument.

Random errors arise from random or chance causes, and must be treated using statistical methods. For example, suppose it is required to measure accurately the mass of an object using a sensitive balance. After the object has been weighed several times, the readings will be found to form a group around a mean value. This may, for example, be because the currents of air in the room containing the balance have disturbed the mechanism during weighing. To specify the precision of the mean value obtained from the group of readings, some indication of the scatter of the readings around the mean is required. To indicate the size of a random error the standard deviation about the mean is specified.

Random Errors

If a large number of readings are taken of a process in which the errors are truly random, the distribution of the values about the mean is Gaussian as shown in figure 1.3. Generally speaking, at least 25 readings are needed before a Gaussian distribution may be assumed. The value of the Gaussian distribution lies in the fact that, when a large enough number of readings has been taken, one can say that 68 per cent of the readings will lie within ±1 standard deviation of the mean and 95 per cent within ±2 standard deviations. In general, if n measurements $x_1, x_2, \ldots, x_n$ of a physical parameter are taken under the same conditions, then the best estimate of the value of the parameter under those conditions is the mean $\bar{x}$ given by

$$\bar{x} = \frac{\sum_{i=1}^{n} x_i}{n} \tag{1.1}$$

The standard deviation s of the set of readings (referred to as a sample by statisticians) is given by

$$s = \sqrt{\left(\frac{\sum_{i=1}^{n} d_i^2}{n}\right)} \tag{1.2}$$

where the values $d_1, \ldots, d_n$ are defined as the difference between the individual readings and the mean, that is

$$d_n = x_n - \bar{x}$$

For a random process (which has a Gaussian distribution) the bounds ±1s and ±2s show how far an individual reading is likely to be from the true value. The normal procedure is to take several readings and find their mean. How far this sample mean is from the true value is of much more concern than the uncertainty of any individual reading. It should be emphasised that there is no way of knowing what the true value is, since an infinite number of readings would be required to calculate it. Obviously, the extent to which the sample mean departs from the true value depends not only on the spread of the individual values (the standard deviation), but also on the number of readings taken. However, it is possible to specify the probability that the sample mean lies within a specific range of the true value. This range is known as the standard error on the mean, s_m, and is calculated from

$$s_m = \frac{s}{(n-1)^{1/2}} \tag{1.3}$$

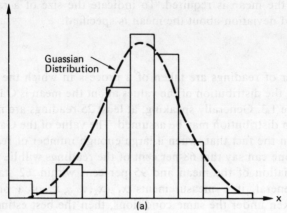
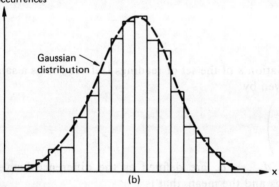
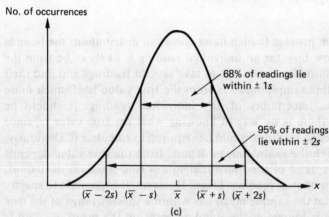

Figure 1.3 Random data and the Gaussian distribution. (a) If > ~25 readings are plotted as a histogram, the result may be approximated to a Gaussian distribution. (b) If more readings are taken and a finer resolution used, a closer approximation is achieved. (c) 68 per cent of data points lie within ±1 standard deviation of the mean in a Gaussian distribution. 95 per cent of the data lies within ±2 standard deviations.

The probability that the mean of a given sample, $\bar{x}$, lies within $\pm s_m$ of the true value is 68 per cent. The probability that $\bar{x}$ lies within $\pm 2s_m$ of the true value is 95 per cent. The standard error s_m is therefore an estimate of how close the mean of a sample set of values $\bar{x}$ is to the true value.

Systematic Errors

From equation (1.3) it can be seen that by taking a large enough number of readings the random error on a measurement may be made as small as desired. However, when a systematic error occurs all the measurements are systematically shifted in one direction, and obviously the process of taking a number of readings and finding a mean value will not improve the accuracy of the measurement. Figure 1.4 shows the spread of readings caused by a random error, and also how the randomly-distributed readings are shifted by a systematic error so that the mean value itself is in error. The terms

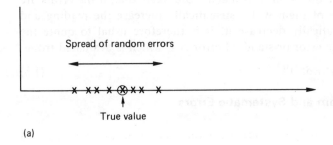

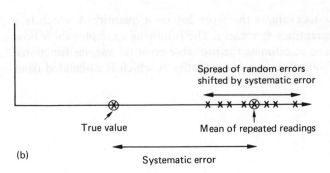

Figure 1.4 The effects of systematic and random errors: (a) random errors; (b) random and systematic errors combined.

'accurate' and 'precise' are used of this situation—a measurement is said to be accurate if the systematic error is small, and precise if the random error is small.

In the introduction to this chapter a zero offset error was given as an example of a systematic error. Another example is the case of a clock that runs either too slowly or too fast. It is impossible to give a complete list of all the possible systematic errors, as each instrumentation system is liable to its own particular hazards. Systematic errors are cumulative, so if for example the desired result A is a function of the quantities x, y and z, that is

$$A = f(x, y, z)$$

then the maximum value of the systematic error ΔA is

$$\Delta A = (\delta x + \delta y + \delta z) \tag{1.4}$$

where δx, δy, and δz are the errors in x, y and z. However, this is usually rather a pessimistic approach, since it requires that all the systematic errors operate in the same direction. It is much more likely that, if the errors are independent, some of them will systematically increase the reading and others will systematically decrease it. It is therefore usual to quote the probable systematic error or standard error ΔA, which is evaluated from

$$\Delta A = (\delta x^2 + \delta y^2 + \delta z^2)^{1/2} \tag{1.5}$$

Combining Random and Systematic Errors

If, as before, we are trying to measure a value $A = f(x, y, z)$ and x, y and z have associated errors δx, δy and δz, which are assumed to be unrelated random or systematic errors, then the resulting standard error ΔA is

$$\Delta A = (\delta x^2 + \delta y^2 + \delta z^2)^{1/2} \tag{1.6}$$

Errors on Sums, Differences, Products and Exponentials

We have seen how to evaluate the error ΔA on a quantity A which is a function of other quantities x, y and z. The following examples show how equation (1.6) is used to estimate the probable error for various functions.

Suppose you want to measure a quantity A which is calculated from the expression

$$A = x + y$$

Equation (1.6) gives the error ΔA as

$$\Delta A = (\delta x^2 + \delta y^2)^{1/2}$$

where x and y are measured quantities with the associated errors δx and δy.

If the quantity A is a difference, that is

$$A = x - y$$

the error is still found from equation (1.6). Although y is negative, the error δy is as likely to be positive as negative. Thus the resultant error is again found from the square root of the sum of the errors squared:

$$\Delta A = (\delta x^2 + \delta y^2)^{1/2}$$

If A is a product of the measured quantities, that is

$$A = xy$$

a slightly different procedure must be followed. Consider the effect of increasing x to $x + \delta x$. Then

$$A + \Delta A = (x + \delta x)y$$
$$= xy + \delta xy$$

that is

$$\Delta A = \delta xy$$

If both sides are divided by A we have

$$\frac{\Delta A}{A} = \frac{\delta xy}{A}$$

$$\therefore \quad \frac{\Delta A}{A} = \frac{\delta xy}{xy}$$

$$\therefore \quad \frac{\Delta A}{A} = \frac{\delta x}{x}$$

Thus, a fractional increase in either x or y produces the same fractional increase in A. The error ΔA is once again evaluated from equation (1.6), but this time fractional errors must be used. We therefore have

$$\frac{\Delta A}{A} = \left\{ \left(\frac{\delta x}{x}\right)^2 + \left(\frac{\delta y}{y}\right)^2 \right\}^{1/2}$$

If there are more than two factors involved the same rule applies, and the error in A is calculated from the square root of the sum of the squares of the fractional errors. Expressed mathematically, if $A = x_1 x_2 x_3 \ldots x_n$, then

$$\frac{\Delta A}{A} = \left\{ \sum_{i=1}^{n} \left(\frac{\delta x_i}{x_i}\right)^2 \right\}^{1/2} \tag{1.7}$$

If A is a ratio, that is $A = x/y$, the procedure to be followed can once again be deduced by considering the effect of increasing x to $x + \delta x$:

$$A + \Delta A = \frac{x + \delta x}{y}$$

$$\therefore \quad A + \Delta A = \frac{x}{y} + \frac{\delta x}{y}$$

$$\therefore \quad \Delta A = \frac{\delta x}{y}$$

$$\therefore \quad \frac{\Delta A}{A} = \frac{\delta x}{y} \cdot \frac{y}{x}$$

that is

$$\frac{\Delta A}{A} = \frac{\delta x}{x}$$

Once again we see that the effect of a fractional change in x or y is to produce the same fractional change in A. Equation (1.7) is therefore again used to evaluate the error in A.

Other Error Analysis Techniques

In the earlier analysis of random errors the assumption was made that the random deviations from the true value were due to chance, and approximated to a Gaussian distribution. So-called 'goodness-of-fit' tests are available to ascertain whether this assumption is correct. Tests are also available to enable the user to decide whether a difference between two sets of observations, such as a change in the mean value or standard deviation, is significant or merely chance. These tests are known as significance tests. There are alternatives to the Gaussian distribution which may be used in special cases, for instance when the distribution of a set of readings is biased in one direction, or when some measurements are to be given more weight than others. Details of these and many other statistical techniques are given in the works listed in the bibliography (at the end of the book) for this chapter.

Chapter 2
Sensors and Transducers

INTRODUCTION

In chapter 1 we saw that the first two components of an instrumentation system are usually a primary sensing element such as a spring, which is mechanical in nature, and a transducer. The transducer has the function of converting the input to an electrical form for convenient transmission, data storage or analysis. In many pressure transducers for example, the pressure being measured is made to deflect a mechanical element such as a diaphragm, bellows or Bourdon tube, and a displacement transducer is used to convert the deflection into an electrical signal.

The aim of this chapter is to describe some of the physical effects that can be used to construct sensing elements and transducers, and to indicate the means by which measurements of a number of engineering quantities may best be made.

The three most important and fundamental measurements are those of displacement, temperature and radiation. Almost all sensors rely on measurements of one of these. For example, one form of microphone works by applying acoustic pressure waves to an elastic diaphragm and measuring the resulting deflections with a capacitive displacement transducer.

It should be emphasised that not all the sensing techniques described in this chapter are available as commercial devices. For example, capacitive displacement sensors are often constructed to suit the needs of a particular application. The construction of accelerometers on the other hand is such that they are invariably obtained from a specialist manufacturer.

DISPLACEMENT SENSING

Measurements of the displacement of an object are of fundamental importance in experimental science, and are the basis of measurements of velocity, acceleration, strain, and (by the use of springs) force and pressure. Either translational or rotational displacement measurements may be needed. The principles underlying the operation of many displacement sensors are common to both linear and rotary types. For this reason these two types of measurement have not been treated separately.

The Linear Variable Differential Transformer (LVDT)

Probably the most common sensor used for displacement measuring is the Linear Variable Differential Transformer or LVDT. This device is basically a transformer, in which the coupling between the primary and secondary coil depends upon the position of a movable ferromagnetic core. The core is usually mounted in a precision linear bearing, and with this arrangement the friction and wear are negligible. There are several designs, but in the most common the secondary is split into two halves, wound in opposition and positioned symmetrically on both sides of the centre of the transducer as shown in figure 2.1. The coils all have the same length L, and the ferromagnetic core has length $2L$. Very precise coil winding is required if the two halves of the secondary are to be matched. Failure to achieve this means that the impedance of the two halves of the secondary will differ, and an unwanted quadrature output will result when the device is in the balance position.

The primary is excited by an alternating current, and voltages V_1 and V_2 are induced in the secondary coils. Since the two halves of the secondary are wound in opposition and connected in series, the output is

$$V_{out} = V_1 - V_2 \tag{2.1}$$

When the core is in its central position it protrudes equally into the two halves of the secondary, and the output is in theory zero. In practice, slight mismatches between the secondary coils usually mean that a small output remains with the core in this position. If the core is moved away from the centre in either direction the output rises linearly as shown in figure 2.1, with 0° phase in one direction and 180° phase in the other. Thus, the magnitude of the output indicates the size of the displacement, and the phase of the output its direction.

Commercially-available LVDTs have strokes from around ±0.5 mm up to ±25 cm or more. The linearity is usually about ±0.25 per cent. The sensitivity depends on the size of the excitation voltage and the stroke, but typically is in the range 0.1 V/cm to 50 mV/μm.

The mass of the movable core in an LVDT is quite small, from 0.1 g in a small device up to around 10 g in a long-stroke LVDT. The maximum practical excitation frequency is usually about 20 kHz. This, combined with the mass of the moving core, restricts the range of motion frequencies that

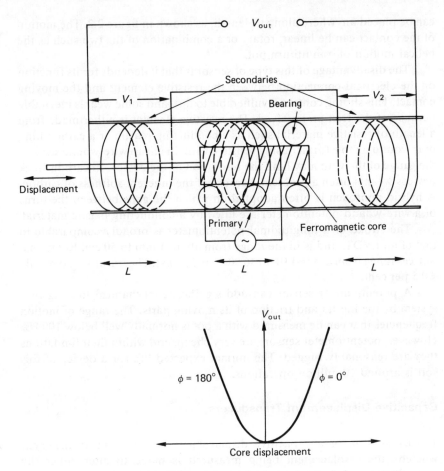

Figure 2.1 Linear variable differential transformer and output characteristic.

can be measured to a maximum of about 2 kHz. However, for many mechanical systems this is perfectly adequate.

Rotary LVDTs operate on exactly the same principle as the linear versions. A linearity of ±1 per cent for a travel of around 50° is common, with a sensitivity of the order of 20 mV/degree.

Potentiometer Displacement Transducers

Variable resistances or potentiometers are used as the basis of a number of cheap displacement sensors. The simple rotary potentiometer is probably the most common type.

Reduced to its essentials, a resistive potentiometer (usually called a 'pot') consists of a resistance element arranged so that a movable contact can be placed anywhere along its length, as shown in figure 2.2. The motion of the contact can be linear, rotary or a combination of the two such as the helical motion of a multiturn pot.

The disadvantage of this type of sensor is that it depends for its function on the electrical connection between the resistive element and the moving contact. This sliding contact is vulnerable to dirt, and some wear is inevitable because of the friction involved. The resistive element is either made from a length of resistive material such as a conducting plastic or a carbon film, or is wound on to a former using resistive wire. In cheap wire-wound devices the gauge of the wire may limit the resolution of the device. In carbon-track devices wear is often a problem, because of the inherent softness of carbon. A hybrid approach is often adopted, in which the gaps between the turns of a wire-wound potentiometer are filled by a conducting plastic material.

The stroke of a longitudinal potentiometer is broadly comparable to that of an LVDT, and is in the range from about 1 mm to 50 cm. Rotational devices range from around 10° to 36 000° (100 turns). The linearity is typically ±0.5 per cent.

A potentiometer sensor can add significant mechanical loading to a system by the inertia and friction of its moving parts. The range of motion frequencies that can be measured with a pot is normally well below 100 Hz. However, potentiometer sensors are very cheap, and within their limitations they are reasonably rugged. The normal expected life for a device of this sort is around 10 million operations.

Capacitive Displacement Transducers

The usual form of a capacitive displacement sensor is an arrangement whereby the displacement being measured is made to alter either the area of facing conductor plates or the separation of the plates as shown in figure 2.3.

The capacitance C of a parallel plate capacitor, area A and separation d, with the space between the plates filled by a dielectric of relative permit-

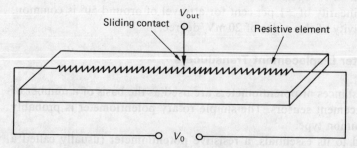

Figure 2.2 *Resistive displacement sensor.*

SENSORS AND TRANSDUCERS 17

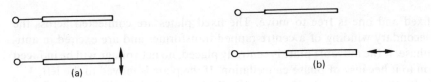

Figure 2.3 *Capacitive displacement sensors: (a) variable separation; (b) variable area.*

tivity ε_r, is

$$C = \frac{\varepsilon_r \varepsilon_0 A}{d} \qquad (2.2)$$

(where ε_0 is the permittivity of free space).

From equation (2.2) it can be seen that the value of C may be changed by altering the spacing between the plates d, the facing area of the plates A, or the relative permittivity of the dielectric between the plates ε_r. However, because of the practical difficulties involved, variable permittivity capacitive transducers are not often used.

Variable-Area Capacitive Sensors

Capacitive sensors which use a pair of plates (such as the one shown in figure 2.3) give a magnitude output. If it is required to measure not only the size of a displacement but also its direction, an arrangement such as that shown in figure 2.4 can be used, where three plates are arranged to give a differential output. Figure 2.4 shows a variable-area capacitive sensor made from three identical plates each of length L. Two of the plates are

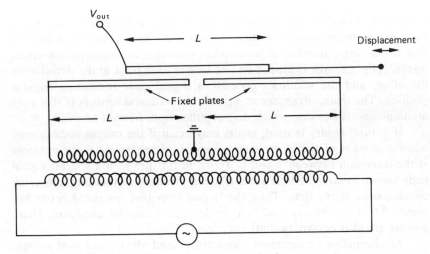

Figure 2.4 *Three-plate variable-area capacitive displacement sensor.*

fixed and one is free to move. The fixed plates are connected across the secondary winding of a centre-tapped transformer and are excited in anti-phase. If the movable plate is centrally placed, no net voltage will be induced on to it because of phase cancellation. If the plate is moved to the left, V_{out} will increase in proportion to the displacement and will have the same phase as the left-hand fixed plate. If the plate is moved to the right, V_{out} will have the same phase as the right-hand plate. Thus, the magnitude of V_{out} indicates the size of the displacement, and the phase of V_{out} its direction. The range over which the device operates is the plate length L.

The voltage output from a capacitive sensor is normally measured by means of a charge amplifier. This is because the impedance of a capacitive sensor $(-j/\omega C)$ is reactive, and a charge amplifier is necessary to ensure that the output signal is independent of frequency.

Variable-Separation Capacitive Sensors

An arrangement which makes use of a variable-separation capacitive sensor is shown in figure 2.5. As before, the device consists of three identical capacitor plates, but this time the central plate is constrained so that it moves perpendicularly between the two fixed plates. The output from the sensor is zero when the movable plate is in the central position, and increases when it is moved to the left or right. Once again the phase of the output indicates the direction of movement. The range of measurement is twice the plate separation d.

Optical Displacement Sensors

Longitudinal and angular motion are often measured using optical methods. For angular measurements optical encoders are used. These consist of a disk containing a number of transparent windows, arranged in concentric tracks. A light source is placed on one side of each track and a detector on the other, and the windows generate a digital code indicating angular position. The main advantage of optical displacement sensors is that they are non-contact devices, which do not suffer from friction or wear.

If natural binary is used, errors may occur if the output code is read when a number of tracks are simultaneously changing state, for example at the transition between 7 and 8. To avoid this problem a special digital code known as Gray code is used, which is arranged so that only one track changes state at any time. Thus, the largest error that can occur is one bit. Figure 2.6 shows binary and Gray code optical angular encoders. These sensors are also known as shaft encoders.

An alternative arrangement is sometimes used which consists of a single binary track and a pair of photocells. This arrangement is shown in figure

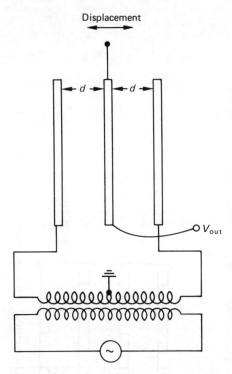

Figure 2.5 Three-plate variable-separation capacitive displacement sensor.

2.6(b), and is referred to as an incremental angular encoder. Since a single track is used, the device may be manufactured cheaply in slotted form rather than as a pattern of opaque and transparent zones. The slots are detected by a pair of photocells positioned one quarter of a slot width apart, as shown in figure 2.6(b). With this arrangement the phase of the two signals indicates the direction of rotation, and the frequency of the pulse train from one photocell gives the rotation rate.

Optical gratings are used for longitudinal displacement measurement as shown in figure 2.7. Figure 2.7(a) shows a grating measurement system consisting of a fixed and a movable grating. When the movable part is displaced the whole field goes dark or light as the lines go in and out of phase, and this change is detected by a photocell. The output is fed to a counting circuit and the count indicates the magnitude (but not the direction) of a displacement.

Figure 2.7(b) shows a system where a phase difference of 90° is introduced between the two halves of the moving grating. The effect of this is to produce a quadrature output which indicates the direction as well as the magnitude of a displacement.

20 INSTRUMENTATION FOR ENGINEERS

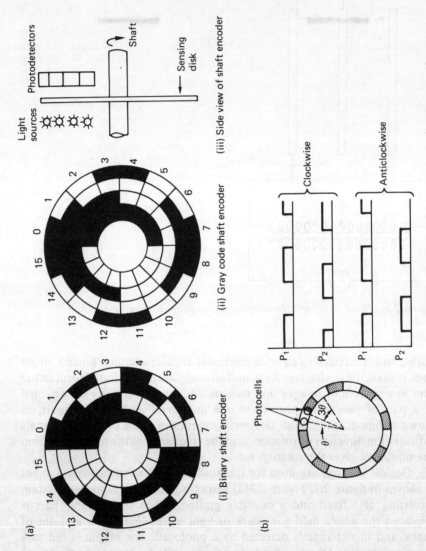

Figure 2.6 (a) Optical shaft encoders. (b) Incremental angular encoder and output waveforms.

SENSORS AND TRANSDUCERS 21

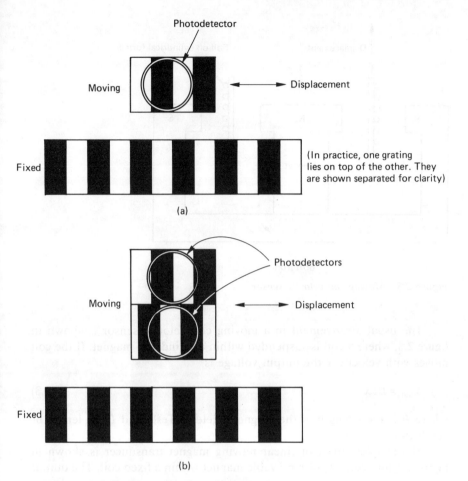

Figure 2.7 Grating systems for displacement sensing: (a) grating system for displacement measuring—magnitude only; (b) grating system for measuring magnitude and direction of displacement.

VELOCITY SENSING

Longitudinal Velocity Sensing

If there is relative motion between an electrical conductor and a magnetic field such that the conductor cuts the field lines, a voltage will be generated across the ends of the conductor. This is the principle of the electric generator, and it is used in a number of velocity sensors. Magnetic velocity sensors can be divided into two types, according to whether a moving coil or a moving magnet is used.

22 INSTRUMENTATION FOR ENGINEERS

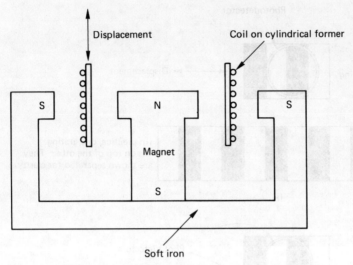

Figure 2.8 Moving coil velocity sensor.

The usual arrangement in a moving coil velocity sensor is shown in figure 2.8, where a coil is suspended within a cylindrical magnet. If the coil moves with velocity $\dot{x}$ the output voltage is

$$V_{out} = BL\dot{x} \tag{2.3}$$

where B is the strength of the magnetic field in Tesla and L the length of conductor cut by the flux.

The simplest form of linear moving magnet transducer is shown in figure 2.9, and consists of a movable magnet within a fixed coil. The output

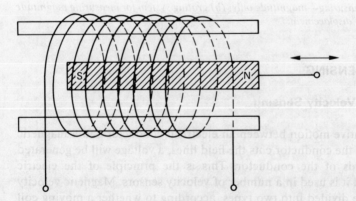

Figure 2.9 Moving magnet velocity sensor.

from this type of sensor can be markedly non-linear however, and so is not often used.

Rotational Velocity Sensing

The tachometer is the rotational equivalent of the moving coil velocity sensor. In a DC tachometer the magnetic field is generated by a permanent magnet, and a coil is rotated between the poles. The induced DC voltage is proportional to velocity, and is obtained from the rotating coil by means of slip rings. (Slip rings are highly polished metal disks or cylinders, which allow signals to be transmitted between the rotating and fixed parts of a machine by means of a sliding contact.) To ensure that the DC output is reasonably smooth, a number of magnetic poles are placed around the coil. However, as the number of poles must be finite some ripple is unavoidable. In addition, DC tachometers can suffer from noise due to imperfections in the electrical contact at the slip rings.

AC tachometers are in general less noisy than the DC type since slip rings are unnecessary, and do not suffer from ripple effects. An AC tachometer consists of a solid rotating conducting cylinder, arranged orthogonally with a pair of coils as shown in figure 2.10. One coil is sinusoidally excited at constant frequency, and because of eddy current effects within the rotating core an AC voltage is induced in the second coil.

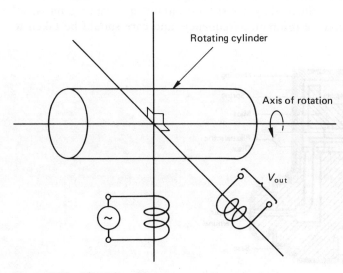

Figure 2.10   The AC tachometer.

The magnitude of the voltage in the second coil is proportional to the rotation rate.

ACCELERATION SENSING

Most acceleration sensors or accelerometers use the piezoelectric effect. This is a phenomenon exhibited by certain crystalline and ceramic materials, in which a potential difference appears across opposite faces as a result of mechanical deformation. The piezoelectric effect is reversible, in that if a voltage is applied across opposite faces of the material it changes its dimensions. Piezoelectric materials form the basis of a number of force and displacement sensors. Since force is directly related to acceleration (Newton's second law), piezoelectric materials are widely used to construct accelerometers by mounting a known mass on a disk of piezoelectric material.

The construction of a common type of accelerometer is shown in figure 2.11. The sensing element consists of a mass resting on a number of piezoelectric disks. The mass and the disks are held in compression by a pre-loading spring, and the whole assembly is sealed into a housing with a thick metal base which can be fixed to the object being investigated. When the accelerometer is subjected to acceleration the mass exerts a force on the disks, and because of the piezoelectric effect a variable charge is developed which is proportional to the force and therefore to the acceleration of the mass.

Accelerometers are used for measuring both shock loadings and continuous vibrations. Since they consist essentially of a spring mass, all accelerometers have a resonance frequency, and care should be taken to

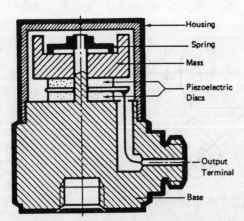

Figure 2.11 Piezoelectric accelerometer (courtesy of Brüel and Kjaer).

ensure that all the frequencies of any vibration being measured fall well below this resonance frequency, so that the acceleration of the mass is equal to that of the whole transducer. This is particularly important for measurements of transient accelerations, since the spectrum of a transient can contain very high frequencies as discussed in chapter 7.

Since the output from an accelerometer is a charge rather than a voltage, charge amplifiers must be used to condition the signal. The electrical connection between an accelerometer and a charge amplifier should be kept as short as possible to reduce noise, and the special low-loss cables used must be clamped to minimise the effects of noise introduced by mechanical motion.

STRAIN MEASUREMENT

The strain in a material is the amount of mechanical deformation suffered by that material due to the action of a load. Strain is a vector quantity in that it has both magnitude and direction, and it is measured using strain gauges which have to be oriented in the direction of measurement. Since strain determination is essentially a measurement of displacement, it is appropriate that strain sensors are discussed in this section.

Early methods of strain measurement were mechanically based, and used systems of levers to magnify the small extensions due to a strain until they were large enough to move a pointer against a calibrated scale. Mechanical strain gauges with a pointer indication are only useful for static strain measurements. For use with dynamic loads scratch gauges can be used, in which the system of magnifying levers actuates a stylus which 'writes' a trace on to a revolving foil or glass disk actuated by clockwork. Mechanical strain gauges are still sometimes used, their great advantage being that they require no external power supply. However, they are generally larger than the much more common electrical resistance strain gauge, and since strain is a property of a point in a material, the larger the gauge the greater the error in strain measurement.

The electrical resistance strain gauge (which from now on we shall simply refer to as a strain gauge) was developed in the 1930s, and the early versions consisted of a length of wire glued to the test object. Changes in length (strains) on the surface of the test object are transferred to the wire, and cause alterations in the resistance of the wire. The resistance changes can be very accurately measured by means of a bridge circuit (see chapter 3). Modern strain gauges are often made by etching a thin metal foil rather than using a wire, but the principle of operation is identical in each case. In recent years semiconductor strain gauges have become available, which are many times more sensitive than wire or foil gauges. We shall deal with the two types separately.

Wire and Foil Gauges

To ensure that the resistance changes are as large as possible a long gauge length must be used. However, it is also necessary to make the gauge occupy as small an area as possible, so that the measurement approximates to point strain determination. To achieve these two objectives the conductor (foil or wire) is normally folded into a grid pattern as shown in figure 2.12. The etching process used to manufacture foil gauges can be made to produce complex shapes for special purposes, such as the examples shown in figure 2.13.

The change in resistance of a gauge is related to the change in gauge length (the strain) by the gauge factor k:

$$k = \frac{\delta R/R}{\delta L/L} = \frac{\delta R}{\varepsilon R} \tag{2.4}$$

where

R = original gauge resistance (that is, without strain)
δR = change in gauge resistance
L = gauge length
δL = change in gauge length
ε = strain.

The conductor used to manufacture a strain gauge should have as high a gauge factor as possible, so that small strains cause large changes in resistance. In addition, the gauge factor must be linear—in other words, multiples of a given extension must produce the same multiple of the resistance change. The gauge factor for a wire or foil gauge is usually about 2. The exact value for a given gauge is supplied with the gauge, and is normally quoted to two decimal places. The following values (courtesy of Brüel and Kjaer) are for typical commercially available wire and foil gauges:

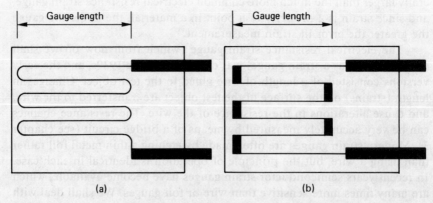

Figure 2.12 Wire and foil strain gauges: (a) wire gauge; (b) foil gauge.

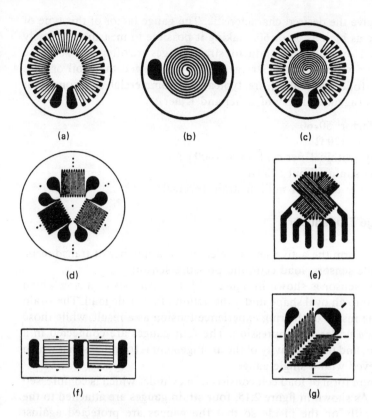

Figure 2.13 Special-purpose strain gauges: (a) radial gauge; (b) tangential gauge; (c) combined radial/tangential gauge; (d) delta rosette; (e) stacked 60° rosette; (f) two-element 90° grid; (g) herringbone grid (courtesy of Brüel and Kjaer).

Gauge factor: approximately 2

Gauge resistance: standardised values of 120 Ω, 350 Ω, 600 Ω and 1000 Ω are used

Linearity: usually within 0.1 per cent up to 4000 $\mu\varepsilon$, and within 1 per cent up to 10 000 $\mu\varepsilon$

Breaking strain: about 25 000 $\mu\varepsilon$

Fatigue life: up to 10 million strain reversals

Temperature compensation: gauges can be obtained with coefficients of thermal expansion that match general-purpose steels, stainless steels and aluminium alloys

Semiconductor Strain Gauges

Semiconductor strain gauges consist of a strip of semiconductor material such as silicon or germanium, which is doped with a controlled amount of impurity to give the desired characteristic. The gauge factor of this type of gauge can be as high as 50 or 60, making it possible to measure extremely small strains. However, semiconductor strain gauges are often very sensitive to temperature variations, and are generally less rugged than foil or wire gauges. The following values are typical for commercial semiconductor strain gauges (again courtesy of Brüel and Kjaer):

Gauge factor: 50–60
Resistance: 120 Ω
Linearity: within 1 per cent up to 1000 $\mu\varepsilon$
Breaking strain: about 5000 $\mu\varepsilon$
Fatigue life: around 1 million strain reversals

Strain Gauge Transducers

Strain gauges form the active sensing element in a number of transducers, notably tensile sensors, load cells and pressure sensors.

A tensile sensor is shown in figure 2.14, and consists of a ring which is deformed into an oval shape under the action of a tensile load. The strain gauges on the inside of the ring experience tension as a result, while those on the outside undergo compression. The four gauges are connected in a bridge circuit, and the sensitivity of the arrangement is four times that which can be achieved with a single gauge.

A common form of load cell consists of a cylinder which is compressed by the load. As shown in figure 2.15, four strain gauges are attached to the cylinder, usually on the inside so that the gauges are protected against damage by abrasion. Two of the gauges are placed so that they measure compression in the cylinder walls, and two so that they measure the resulting lateral deformation due to Poisson's ratio. Poisson's ratio is around 0.3 for most engineering metals, so this arrangement is 2.6 times more sensitive than a single gauge as explained in chapter 3.

One form of pressure transducer consists of a thin circular diaphragm which is deformed by the action of pressure. A pair of gauges (one on each

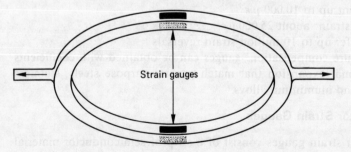

Figure 2.14   Strain gauge tensile sensor.

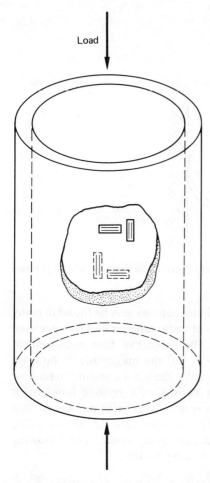

Figure 2.15 Strain gauges inside tube used as load cell.

side) is mounted at the centre of the diaphragm, and the remaining pair mounted in the comparatively unstressed area at the edge of the diaphragm as shown in figure 2.16. The unstressed gauges are known as dummy gauges, and are included in the bridge circuit to provide automatic temperature compensation as shown in chapter 3. The gauges at the centre of the diaphragm give an output which is proportional to pressure.

FLOW SENSORS

Flow sensors are important in a number of applications, particularly those where the measurements are used to control a process. Examples are

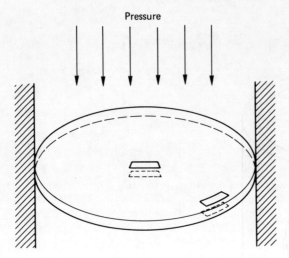

Figure 2.16 A strain-gauged diaphragm, constrained at its edges, used as a pressure gauge.

common in manufacturing industry, and flow sensors may be found in many modern motor cars where measurements of instantaneous fuel flow are used by the engine management computer. The subject of flow measurement includes vector flow, where the direction and the magnitude of flow are sensed, volume flow and mass flow rate. We shall not consider mass flow rate sensors here, since they generally consist of a volume flow sensor followed by a density measurement. There are a very large number of flow sensors available which space precludes us from considering, including some very accurate non-invasive techniques using lasers. The following descriptions are of the most common low-cost devices.

Vector Flow Transducers

The two most common methods of measuring vector flow are by means of hot-wire anemometers and by pitot tubes. A hot-wire anemometer consists of an electrically heated wire probe which is placed in the moving fluid. The rate at which thermal energy is lost from the wire depends on both the rate and direction of flow. The wire has a known resistance/temperature characteristic, and is made to form one arm of a bridge circuit (see chapter 3). Either the heating current is maintained at a constant value, or a feedback system is used to keep the temperature of the wire constant. This second approach has the advantage that the heating current is then proportional to the flow.

Figure 2.17 shows the arrangement used to measure vector flow with a pitot tube. The 'tube' actually consists of two concentric tubes, the inner tube having its open end facing the oncoming flow. The outer tube is closed

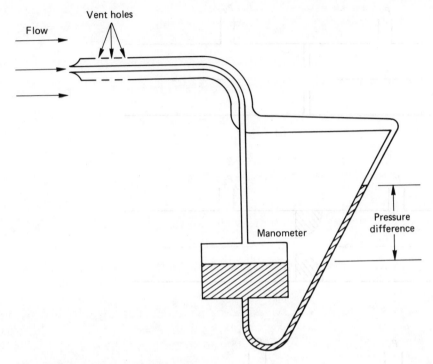

Figure 2.17 Pitot tube flow sensor.

at its end but has a number of holes in the walls. Both tubes contain the same fluid as the flowing medium. The pressure in the outer tube is the static pressure in the fluid. However, the pressure in the inner tube is the sum of the static pressure and a pressure due to the impact of the flow on the stationary fluid in the tube. Thus, the flow generates a pressure differential across the two parts of the pitot tube, which is measured by a manometer as shown in figure 2.17. Obviously, the magnitude of the pressure differential depends on both the flow rate and direction.

Volume Flow Sensors

Most flow sensors are modifiers which partly restrict the flow by means of plates or nozzles to produce a pressure drop which can be measured. Some typical devices are shown in figure 2.18. In each case the volume flow rate is proportional to the square root of the pressure difference.

The drawback of flow sensors such as those illustrated in figure 2.18 is that they disturb the flow by their presence. Turbine flow sensors are more satisfactory, in that the amount of disturbance they cause is smaller. A turbine flow sensor consists of a small vaned device which is placed in the flow. The moving fluid causes the turbine to rotate, and the rotation

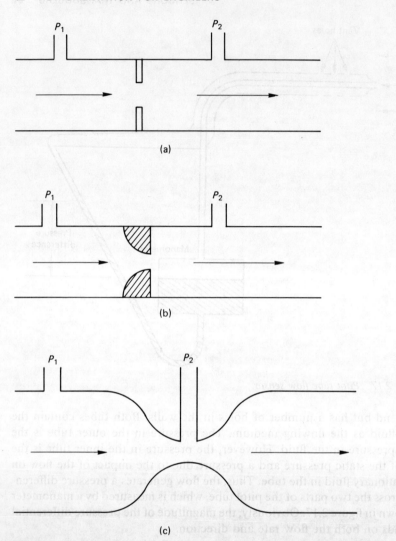

Figure 2.18 Volume flow sensors: (a) orifice plate; (b) nozzle; (c) venturi.

rate is usually sensed externally to the flow (by, for example, a Hall effect sensor).

TEMPERATURE SENSORS

There are three main types of temperature sensor:
(a) resistive devices
(b) semiconductor devices
(c) thermoelectric devices.

Semiconductor temperature sensors or thermistors can be considered as a form of resistive sensor. The classical resistance thermometer is made from a metal wire having a highly temperature-dependent resistivity. The main difference between the two types is that the resistance changes obtained from a resistance-wire thermometer are reasonably linear, but as shown by figure 2.19 the behaviour of a thermistor is non-linear. The output may be linearised by analogue circuits, but the advent of microprocessor-based instrumentation systems has meant that non-linearity is no longer much of a disadvantage, and linearisation can easily be carried out by software.

Resistive Temperature Sensors

These transducers consist of non-inductively wound coils of wire. A wire is chosen which has a strongly temperature-dependent resistivity. Platinum, nickel and copper are the most commonly used materials. The coil normally has a resistance of around 100 Ω, and may be encapsulated in the tip of a glass rod which is used as a probe. Resistance thermometers can be used from about $-250°C$ to $+500°C$, and have a sensitivity of about $0.5\ \Omega/°C$. They are normally connected as part of a bridge circuit (see chapter 3).

Semiconductor Temperature Sensors (Thermistors)

A thermistor is a semiconductor with a temperature-dependent resistivity which is an order of magnitude greater than that of a resistance thermometer.

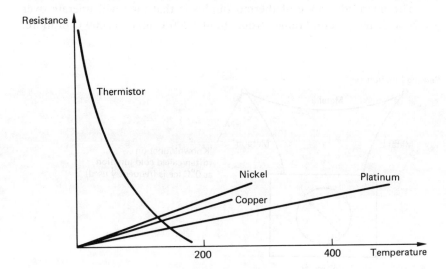

Figure 2.19 Temperature characteristic of metal resistance thermometers and thermistor.

The temperature coefficient of resistance changes by about −5 per cent per °C at room temperature. Normally the temperature coefficient is negative, so that increasing the temperature of a thermistor decreases its resistance.

Thermistors are constructed by compressing and sintering mixtures of nickel, manganese and cobalt oxides into small beads. The size of a thermistor bead can be as small as 0.1 mm in diameter, making extremely localised temperature measurement possible. Thermistors respond to temperature changes very quickly, and commercially available devices can have a response time of a few milliseconds.

The room temperature resistance of a thermistor can be anything between 100 Ω and 1 MΩ. They are generally used within the temperature range from −60°C to +200°C.

Thermoelectric Sensors

The best-known thermoelectric sensor is the thermocouple, which consists of a junction between a pair of wires of dissimilar metals (often platinum and an alloy of rhodium and platinum). Figure 2.20 shows the usual two-junction arrangement, where one junction is attached to the object under thermal investigation and one is maintained at a known temperature (often 0°C using ice). The output from a thermocouple is a small voltage of a few tens of microvolts per °C. High-gain amplification must therefore be used to bring the amplitude of the output signal up to a usable level.

The main advantage of thermocouples is that they will operate over a very wide temperature range, from about −200°C up to +2000°C or more.

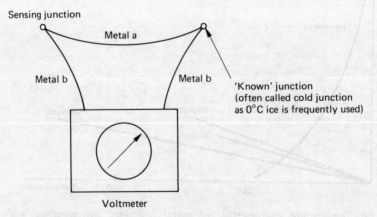

Figure 2.20 Thermocouple temperature sensor.

OPTICAL SENSORS

Sensors which detect electromagnetic radiation in the visible and infrared range are usually referred to as photosensors or photodetectors. Photosensors detect incident radiation in one of two ways. In the first type, normally referred to as thermal photosensors, the radiation is absorbed by an object (which is usually black), and causes a measurable rise in the temperature of the object. The temperature rise is measured by one of the methods described above, often using a thermistor or a thermocouple.

In the second kind of optical sensor, photons of incident radiation are detected by the photoemissive, photovoltaic or photoresistive effects. These detectors are usually called photoelectric sensors.

Thermal Photosensors

Figure 2.21 is a diagram of a thermocouple radiation detector. The screen used to collect radiation forms the hot junction of the thermocouple, and is usually made from gold foil blackened by a coating of carbon. The casing around the device may be evacuated or may be filled with an inert gas, and the radiation is allowed to enter through a window at one end.

The main advantage of thermal photosensors is that their sensitivity is constant for a very wide range of wavelengths of incident radiation. However, the output signal is usually small, often around 10 V per watt of incident radiation. (For the purpose of comparison, the amount of solar power reaching the surface of the Mojave desert in the USA at noon is of the order of 1000 W/m^2.) The frequency response of a thermal photosensor is usually low, a typical device having a time constant of about 20 ms. The poor sensitivity and frequency response mean that thermal photosensors are generally only used for calibration purposes.

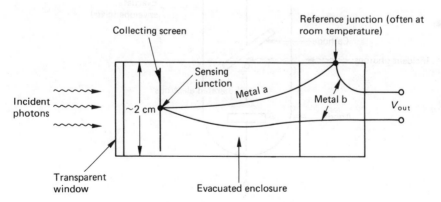

Figure 2.21 *Thermocouple radiation sensor.*

Photoemissive Sensors

The photoelectric effect is a process by which conduction electrons in a metal or other conducting material absorb energy from incident electromagnetic radiation, and escape from the substance.

In a photoemissive sensor an anode and a cathode are arranged inside an evacuated glass envelope as shown in figure 2.22. A voltage is applied across the device, and the electrons produced by the photoelectric effect form a current which can be measured. This current is proportional to the intensity of the incident radiation.

The sensitivity of a photoemissive sensor is fairly high, typically around 10^4 V per watt of incident power. The range of wavelengths over which a photoemissive sensor will work is limited compared with that of a thermal photosensor, and is usually restricted to wavelengths shorter than 10^{-6} m.

Other disadvantages of photoemissive sensors are that they are bulky, fragile and require excitation voltages in excess of 100 V. However, the high sensitivity and speed of response (<10 ns) makes them useful for measuring very low light levels.

Photovoltaic Sensors

Photovoltaic sensors are semiconductor devices in which incident radiation causes the splitting of electron–hole pairs adjacent to a *PN* junction, giving

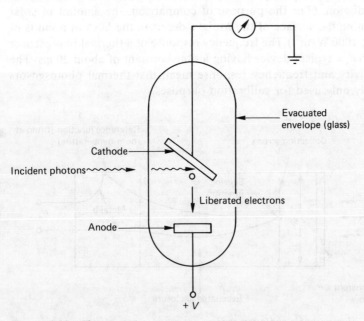

Figure 2.22   Photoemissive sensor.

rise to a current I_L. The device behaves like a semiconductor diode when not illuminated, but with a voltage/current characteristic which moves downward by an amount equal to I_L when the device is illuminated as shown by figure 2.23. If the device is operated under open-circuit conditions, voltage V_{oc} is measured, while if it is operated under short-circuit conditions (for example, when an ammeter is connected across the terminals) current I_L is measured. The device forms the basis of the solar cell, used among other things to provide electrical power for spacecraft.

The sensitivity of a photovoltaic sensor is typically 50 mA (short circuited) or 10^4 V (open circuit) per watt of incident radiation. The device can be operated in voltage or current mode, but is generally used to produce current since the output is then proportional to the intensity of the incident radiation.

Photoresistive or Photoconductive Sensors

These devices are made from semiconductor materials in which photons of incident radiation cause a change in conductivity. The effect of illuminating a sensor of this type is to reduce its resistance. Photoresistive sensors are usually operated as part of a bridge circuit (see chapter 3), and sensitivities up to 10^5 V per watt of incident power can be achieved. However, the sensitivity of this type of sensor is strongly frequency-dependent, falling off

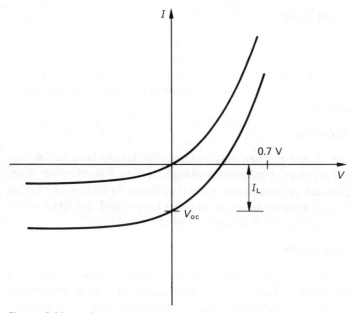

Figure 2.23 *Voltage–current characteristic of a photodiode.*

rapidly below about 10^{14} Hz. The time constant of a photoresistive sensor is of the order of 100 μs. With radiation having stationary spectral characteristics the device is reasonably linear, but if the incident beam has a varying spectral characteristic marked non-linearities can occur.

ACOUSTIC SENSORS

The term acoustics is used here somewhat loosely, and includes ultrasonic compression waves as well as those in the audible range. Acoustic signals are detected by means of microphones. Microphones are constructed in a number of ways, and use some of the phenomena described earlier in this chapter such as the piezoelectric effect, magnetic velocity sensing and variable capacitance.

Ceramic microphones

Ceramic microphones use a disk of a piezoelectric ceramic material, such as lead titanate or lead zirconate, as the transducer. A suitable pre-amplifier is necessary to condition the signal before it can be used, and this is normally mounted close to the microphone to eliminate the possibility of electrical interference. The pressure variations which make up an acoustic wave are transferred to the ceramic disk by means of a diaphragm which forms the primary sensing element.

Capacitor Microphones

An alternative construction is often used in which the acoustic pressure variations are made to deform a diaphragm which forms one plate of a capacitor. Capacitor microphones generally have a better frequency response than the ceramic type, but they require more complicated signal conditioning circuitry.

Electret Microphones

An electret is an almost permanently polarised dielectric material. A diaphragm made of an electret material is placed close to a conducting plate, and acoustic pressure variations cause the diaphragm to vibrate. A varying potential difference appears between the diaphragm and the fixed plate, which is proportional to the displacement of the diaphragm.

Magnetic Microphones

The moving coil velocity sensor described earlier in this chapter is also used to make microphones of high quality. Once again, the sound pressure is used to distort a flexible element such as a diaphragm, which is connected to a moving coil within a magnetic field.

Hydrophones

Microphones used for underwater sound measurement are known as hydrophones. They often use piezoelectric ceramic sensing elements, since these are generally more rugged than other types of microphone.

The important characteristics of microphones are their frequency response, dynamic range, directional properties and ruggedness. It is important to be aware of the limitations of measurements made using a microphone. For example, at high frequencies the response may be limited by the physical presence of the microphone unless special precautions are taken.

It is important to protect microphones from mechanical damage, moisture (except hydrophones, obviously!), and low-frequency turbulence such as that produced by wind. Windshields made from open-cell foams are often used even for indoor measurements, to provide protection against both low-frequency noise and physical damage. In higher-velocity airstreams a streamlined nose cone is often used. For laboratory work microphones are usually calibrated before use by means of a pistonphone, which provides a known acoustic signal at the microphone diaphragm at one or more frequencies.

HALL EFFECT SENSORS

In 1879 the American physicist E. C. Hall discovered that if a metal plate carrying a current is placed in a magnetic field perpendicular to the plate, a voltage appears across the width of the plate as shown in figure 2.24. For

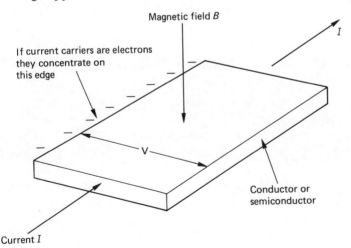

Figure 2.24 Hall effect produces voltage V when a conductor is placed in a magnetic field B.

a conducting plate of thickness t the Hall voltage is given by $V = K_H BI/t$ where K_H is the Hall coefficient, which is a function of the charge mobility and the resistance of the conductor. The Hall voltage is very small in most metals, but can easily be detected in semiconductors. The Hall effect is an illustration of the effect of a magnetic field on a charged particle. The field causes the current carriers to be concentrated at one edge of the conductor, and this bunching causes a voltage to appear across the plate.

Hall effect sensors can be used for current measurement, but are more often used for magnetic field measurement. The voltages produced are of the order of a few microvolts, and the Hall effect is strongly temperature-dependent so the sensitivities obtained are low. A common application of Hall effect sensors is in non-contact measurement of rotation rate. A small permanent magnet is placed on the rotating object, and the output of a nearby Hall effect sensor is a series of pulses, the spacing of which indicates the speed of rotation.

Chapter 3
Signal Conditioning

INTRODUCTION

It is very unusual for a sensor to produce a signal in a form which can be used directly. The amount of energy obtained from a sensor is generally so small that amplification is needed. The required signal may be obscured by electrical noise, in which case filtering will be necessary to recover the data. For practical reasons it may not have been possible to directly measure the quantity of interest, and a related parameter may have been sensed instead which has to be subsequently converted. This often occurs in vibration measurement for example, where the velocity of a vibrating object is measured by using an accelerometer and integrating the signal. (The slightly dubious justification for this process is that accelerometers have a much higher frequency response than most velocity sensors. However, in general it is better to measure a required parameter directly.)

In chapter 1 we saw that there are three fundamental types of sensor or input transducer: modifiers, self-generators and modulators. The output from all three is usually electrical, and a source of external power is necessary in the case of the modulating type. Many modulating sensors rely upon resistance or impedance changes for their function, and these are usually connected as part of a bridge circuit. Bridge circuits (based upon the familiar Wheatstone bridge) are of great importance in instrumentation, and a section of this chapter is devoted to their construction.

In the past, signal conditioning circuits were constructed from discrete components such as bipolar or field effect transistors. (The writer has been told by a retired engineer that once upon a time amplifiers were made from high-voltage devices known as 'valves'. However, no evidence for the truth of this story has been found to date.)

42 INSTRUMENTATION FOR ENGINEERS

Since the advent of cheap and readily available integrated circuits, most signal conditioning has been carried out with the aid of a device called the operational amplifier. With the addition of a few passive components such as resistors and capacitors, the op-amp as it is affectionately known can be made to perform almost all of the required analogue signal conditioning functions. For this reason the circuits described in this and the next chapter are based on op-amps. While it is perfectly possible to construct circuits which perform equivalent functions using discrete semiconductor devices, to do so is nowadays unnecessary and almost unknown.

The topic of filter design is a large one which deserves a chapter to itself, so a discussion of analogue filter circuits has been deferred until the next chapter. This artificial distinction between signal conditioning circuits and filters is made so that sufficient space can be devoted to a subject of fundamental importance to instrumentation engineers.

BRIDGE CIRCUITS

As discussed earlier, many modulating sensors control the flow of electrical energy by changes in resistance or impedance. For example, the change in resistance of a 120 Ω foil strain gauge at 1000 $\mu\varepsilon$ is typically of the order of 0.001 Ω.

Bridges may be excited by an AC or DC power supply. DC bridges are always composed of resistances and are referred to as resistive bridges. An AC bridge can contain resistors, inductors or capacitances, and is known as an impedance bridge.

Resistance (DC) Bridges

The best-known example of a resistive bridge is the Wheatstone bridge, an example of which is shown in figure 3.1. A bridge simply consists of a pair of parallel potential dividers which are connected across an excitation voltage E. When all four resistances have an equal value R the bridge is balanced, that is the voltmeter connected across the centre of the circuit registers zero.

Considering the voltages across each resistor in turn, and using notation based on figure 3.1, we can write:

$$V_{24} = I_{241} R_a = V_{23} = I_{231} R_b$$

and

$$V_{14} = I_{241} R_c = V_{13} = I_{231} R_d$$

so

$$I_{241} R_a = I_{231} R_b \tag{3.1}$$

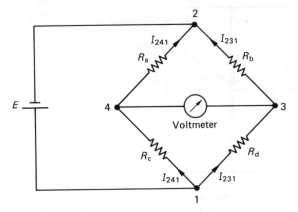

Figure 3.1 Wheatstone (resistive) bridge.

and

$$I_{241} R_c = I_{231} R_d \tag{3.2}$$

If we divide equation (3.1) by (3.2) we obtain:

$$\frac{R_a}{R_c} = \frac{R_b}{R_d} \tag{3.3}$$

Equation (3.3) shows that a change in the resistances on one side of the bridge can be balanced by adjusting the resistance values on the other side of the bridge. It will be demonstrated later that this forms the basis of a method of compensating bridges so that ambient temperature changes do not affect the measurement. Further, equation (3.3) suggests that the sensitivity of a bridge can be enhanced if more than one sensor is included in the measuring circuit.

If one or more of the resistances in a DC bridge are sensors, some means of balancing the bridge before a measurement must be provided. If for example strain is to be measured, the normal sequence of events would be to balance the bridge, apply a load to the test object, and calculate the resulting strain from the out-of-balance voltage produced by the load. The simplest method of balancing a bridge is to make one of the resistors a potentiometer. However, this prevents bridges being constructed in which all four arms are sensors. Figure 3.2 shows an alternative way of balancing a resistor bridge. The advantage of this method is that it can be used regardless of how many fixed resistors and sensors make up the bridge.

The most common resistive sensor is of course the strain gauge. The examples given in this section are all of strain measurement, but it should be emphasised that in general the discussion applies to all resistive sensors.

44 INSTRUMENTATION FOR ENGINEERS

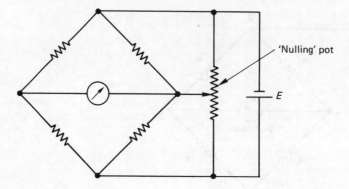

Figure 3.2 Method of balancing a Wheatstone bridge.

It was stated above that a bridge may contain up to four sensors. Bridges with one sensor (and three fixed resistors) are referred to as quarter-bridges. Bridges with two sensors are known as half-bridges, and bridges with four sensing elements are called full bridges. The behaviour of each configuration is analysed in the following sections:

The Quarter-Bridge

If one of the resistances shown in figure 3.1 is a sensor and the other three are fixed, a change in the parameter being sensed will unbalance the bridge. Suppose R_a is a strain gauge with an initial resistance R. When a strain is applied the value of R_a becomes $R + \delta R$, while the other resistances remain unchanged (and equal to R). Then

$$I_{241} = \frac{E}{R_a + R_c} = \frac{E}{2R + \delta R}$$

$$V_{24} = R_a I_{241} = (R + \delta R)\left(\frac{E}{2R + \delta R}\right)$$

$$= \left(\frac{ER + E\delta R}{2R + \delta R}\right)$$

If we consider the other side of the bridge:

$$V_{23} = R_b I_{231} = R_b\left(\frac{E}{R_b + R_d}\right) = \frac{E}{2}$$

So the out-of-balance voltage V across the bridge is

$$V = V_{23} - V_{24} = \left(\frac{E}{2}\right) - \left(\frac{ER + E\delta R}{2R + \delta R}\right)$$

$$= \frac{2ER + E\delta R - 2ER - 2E\delta R}{4R + 2\delta R}$$

If R is small, we can say that $4R + 2\delta R \approx 4R$. Thus, since δR may have either sign we can write

$$V = \pm \frac{E\delta R}{4R} \tag{3.4}$$

Substituting equation (2.4) into (3.4), we find that if R_a is a strain gauge, the strain ε is given by

$$\varepsilon = \frac{4V}{Ek} \tag{3.5}$$

where k is the gauge factor. Equation (3.5) shows that if the excitation voltage E and the gauge factor k are known, the strain may be determined by measuring the out-of-balance voltage.

If a strain gauge is used with a coefficient of thermal expansion which differs from that of the object being measured, it is impossible to say whether a measured strain is due to an applied load, differential thermal expansion or a mixture of the two. However, a feature of bridge circuits is that compensation for thermal effects may be obtained by incorporating a compensating resistor into one of the arms of the bridge. This resistor must have the same temperature characteristic as the sensor, and it is common to use a dummy sensor as compensating resistor. With strain gauges, for example, it is usual to place a gauge identical to that used for sensing on an adjacent, unstressed part of the test object. This dummy gauge is then included in the bridge circuit, and compensates for the effect of any temperature changes on the active or sensing gauge.

The compensating resistor in a quarter bridge is connected in a position adjacent to the sensor. Thus, if for example R_a in figure 3.1 is a strain gauge, the dummy gauge must be R_b or R_c.

Suppose R_a is an active strain gauge, and R_c is a dummy gauge for temperature compensation. Let the effect of a temperature change be to increase the resistance of both R_a and R_c by an amount ΔR. Then

$$I_{241} = \frac{E}{R_a + R_c} = \frac{E}{2R + 2\Delta R}$$

$$V_{24} = R_a I_{241} = (R + \Delta R)\left(\frac{E}{2R + 2\Delta R}\right)$$

that is

$$V_{24} = \frac{E}{2} \tag{3.6}$$

As R_b and R_d are unaffected by the change in temperature, V_{23} is unaltered. The out-of-balance voltage V is

$$V = V_{23} - V_{24} = \frac{E}{2} - \frac{E}{2} = 0 \tag{3.7}$$

46 INSTRUMENTATION FOR ENGINEERS

The effect of any temperature-induced resistance change has been cancelled out, and the bridge is still balanced in the absence of strain. If the active gauge is subjected to a strain and a temperature change while the dummy gauge is subjected to a temperature change alone, the thermal effects cancel out. Any out-of-balance voltage measured across the bridge with this arrangement is due to strain alone.

The Half-Bridge

Figure 3.3 shows a bridge circuit containing two sensors. A bridge containing a pair of sensors in adjacent positions has two main advantages over the single-sensor arrangement discussed in the previous section. First, the sensitivity is enhanced, because the additional sensor gives a larger out-of-balance voltage. Second, the thermal changes to each sensor cancel out, as shown in the previous section.

If we consider as an example the strain-gauged cantilever beam undergoing simple bending shown in figure 3.4, we see that the applied load puts the top surface of the beam into tension and the lower surface into (numerically equal) compression. The resistance of strain gauge R_a becomes $R + \delta R$ under the action of the bending, while that of gauge R_c becomes $R - \delta R$. We may calculate the out-of-balance voltage expected with this arrangement as follows:

$$I_{241} = \frac{E}{R_a + R_c} = \frac{E}{2R}$$

$$V_{24} = R_a I_{241} = (R + \delta R) \cdot \frac{E}{2R} = \frac{ER + E\delta R}{2R}$$

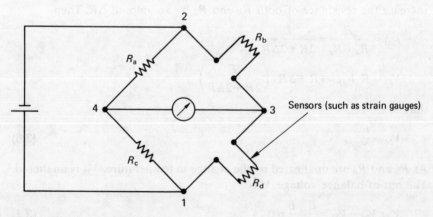

Figure 3.3 The half-bridge.

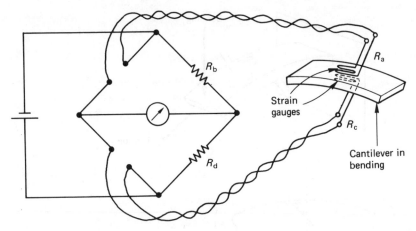

Figure 3.4 Strain gauge in half-bridge connection used to measure cantilever undergoing bending.

We saw earlier that $V_{23} = E/2$, so the out-of-balance voltage V is

$$V = V_{23} - V_{24} = \left(\frac{E}{2}\right) - \left(\frac{ER + E\delta R}{2R}\right)$$

$$= \left(\frac{ER - ER - E\delta R}{2R}\right)$$

$$= \pm \frac{E\delta R}{2R} \qquad (3.8)$$

Comparing equation (3.8) with equation (3.4), we see that a half-bridge has twice the sensitivity of a quarter-bridge.

If the cantilever has a uniform cross-section, its neutral axis will coincide with the centre of the beam. The stress at any point within the beam can then be calculated from the measured surface strains.

If however the cantilever is subjected to a tensile force in addition to bending, as shown in figure 3.5, both strain gauges will be stretched by the same amount. The form of the resulting stress distribution is shown in figure 3.5. Since the active gauges are connected in adjacent arms of the measuring bridge, the extension of each gauge due to the tensile load will cancel out in the same way as a thermal gauge extension. This can be useful if only the bending component is to be measured, but obviously any calculation of total stress made from the strain gauge bridge measurement will in this instance be in error by the amount of the tensile stress.

Figure 3.6 shows a tensile test, in which a pair of orthogonally mounted strain gauges has been applied to the test specimen. With the gauges connected as shown in figure 3.6 the effect of the strain is to increase the

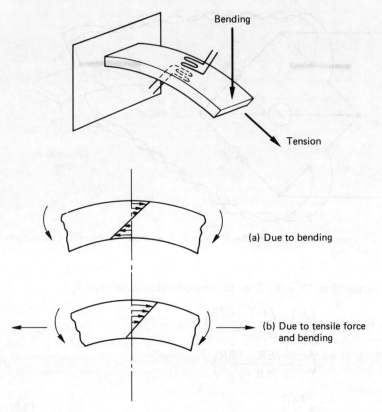

Figure 3.5 Strain-gauged cantilever undergoing tension and bending, with associated stress distributions.

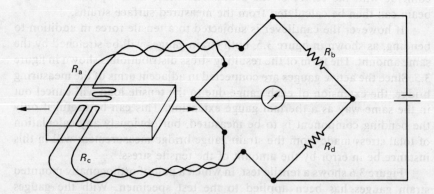

Figure 3.6 Tensile test, half-bridge connection.

resistance of R_a to $R+\delta R$. If a bar such as that of figure 3.6 is subjected to longitudinal tensile stress it will extend in the direction of the stress, and contract in the transverse direction. The transverse strain is proportional to the longitudinal strain, and the constant of proportionality is known as Poisson's ratio ν. For most metals ν is in the range 0.28 to 0.32, with 0.3 being a typical value.

If the test specimen of figure 3.6 has a Poisson's ratio ν, the effect of the tensile stress will be to decrease the value of R_c to $R-\nu\delta R$. The out-of-balance voltage in this case can now be calculated:

$$I_{241} = \frac{E}{R_a + R_c} = \frac{E}{R+\delta R + R - \nu\delta R} = \frac{E}{2R+(1-\nu)\delta R}$$

$$V_{24} = R_a I_{241} = (R+\delta R)\left(\frac{E}{2R+(1-\nu)\delta R}\right)$$

$$= \left(\frac{ER + E\delta R}{2R+(1-\nu)\delta R}\right)$$

Once again, $V_{23} = E/2$ and the out-of-balance voltage V is

$$V = V_{23} - V_{24} = \left(\frac{E}{2}\right) - \left(\frac{ER + E\delta R}{2R+(1-\nu)\delta R}\right) = \frac{\pm(1+\nu)E\delta R}{4R+2(1-\nu)\delta R}$$

If the value of δR is small, $4R+2(1-\nu)\delta R \approx 4R$, so that

$$V = \pm\frac{(1+\nu)E\delta R}{4R} \tag{3.9}$$

Comparing equations (3.9) and (3.4), we see that the sensitivity of a half-bridge used in this way to measure tensile strain is 1.3 times that of a quarter-bridge for $\nu = 0.3$.

In addition to improving the sensitivity by 30 per cent, the arrangement described above also provides automatic temperature compensation. As described earlier, if temperature changes occur during a test and the gauges are not matched to the material under investigation, both gauges will experience the same change in resistance and the thermal effects will cancel out.

The Full Bridge

The foregoing discussion of quarter-bridges and half-bridges has shown that it is possible to enhance the sensitivity obtained from a measuring bridge if more than one active sensor is connected into the circuit. The greatest sensitivity is obtained when all four arms of a bridge contain active sensors. When a full bridge is used it is not usually possible to adjust the resistance of one of the arms of the bridge to balance it, and an arrangement such as that shown in figure 3.2 must be used.

50 INSTRUMENTATION FOR ENGINEERS

With a full strain gauge bridge all four gauges are usually exposed to the same temperature variations. The apparent out-of-balance signal due to thermal effects cancels out as before.

Figure 3.7 again shows a strain-gauged cantilever undergoing bending. When the gauges are connected as shown the following resistance changes occur under the load:

R_a increases to $R + \delta R$

R_b decreases to $R - \delta R$

R_c decreases to $R - \delta R$

R_d increases to $R + \delta R$

Calculating the out-of-balance voltage as before:

$$I_{241} = \frac{E}{R_a + R_c} = \frac{E}{2R}$$

$$V_{24} = R_a I_{241} = (R + \delta R)\left(\frac{E}{2R}\right) = \frac{ER + E\delta R}{2R}$$

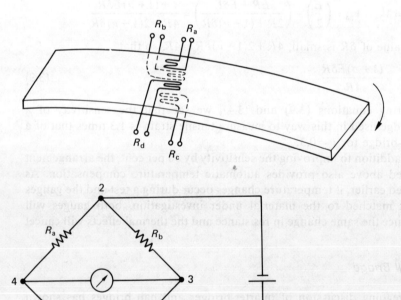

Figure 3.7 Bending test, full bridge connection.

By similar working

$$V_{23} = \frac{ER - E\delta R}{2R}$$

So the out-of-balance voltage V is

$$V = V_{23} - V_{24}$$

$$= \frac{ER - E\delta R - ER - E\delta R}{2R} = \pm\frac{E\delta R}{R} \quad (3.10)$$

By comparing equation (3.10) with (3.4) we see that a full bridge used to measure bending strain has four times the sensitivity of a bridge containing only one strain gauge.

The use of a full bridge in tensile strain measurement gives greater sensitivity than a quarter-bridge or half-bridge. Figure 3.8 shows a tensile test specimen which has been instrumented with four strain gauges, two mounted longitudinally and two in the transverse direction. If the test

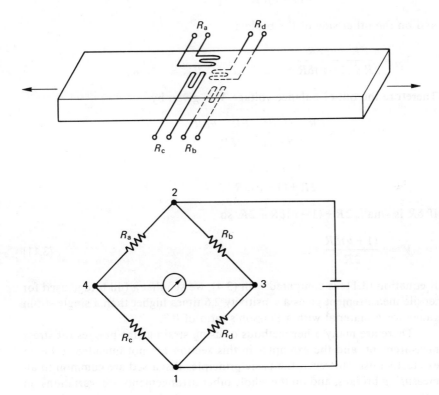

Figure 3.8 Tensile test, full bridge connection.

material has a Poisson's ratio of ν, the following changes in gauge resistance will take place under load:

R_a increases to $R + \delta R$

R_b decreases to $R - \nu\delta R$

R_c decreases to $R - \nu\delta R$

R_d increases to $R + \delta R$

Proceeding as usual to calculate the out-of-balance voltage:

$$I_{241} = \frac{E}{R_a + R_c} = \frac{E}{2R + (1-\nu)\delta R}$$

$$V_{24} = R_a I_{241} = (R + \delta R) \cdot \left(\frac{E}{2R + (1-\nu)\delta R}\right)$$

$$= \frac{ER + E\delta R}{2R + (1-\nu)\delta R}$$

and on the other side of the bridge:

$$V_{23} = \frac{ER - \nu E\delta R}{2R + (1-\nu)\delta R}$$

Therefore the out-of-balance voltage V is given by

$$V = V_{23} - V_{24} = \frac{ER - \nu E\delta R - ER - E\delta R}{2R + (1-\nu)\delta R}$$

$$= \pm \frac{(1+\nu)\delta R}{2R + (1-\nu)\delta R}$$

If δR is small, $2R + (1-\nu)\delta R \approx 2R$, so

$$V = \pm \frac{(1+\nu)\delta R}{2R} \tag{3.11}$$

If equation (3.11) is compared with (3.4), we see that a full bridge used for tensile measurement gives a sensitivity 2.6 times higher than a single strain gauge for a material with a Poisson's ratio of 0.3.

There are many other methods of using strain gauge bridges for stress measurement, and the examples in this section are not intended to be an exhaustive list. However, the basic principles discussed are common to all measuring bridges, and on the whole other arrangements are 'variations on a theme'.

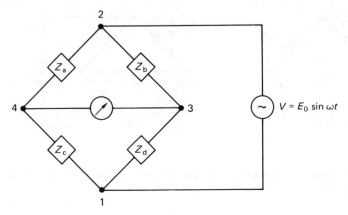

Figure 3.9 AC bridge composed of impedances Z_a, Z_b, Z_c and Z_d, excited by AC voltage $E_0 \sin \omega t$.

AC Bridges

The majority of variable inductance or capacitance sensors are used as part of an AC bridge. AC bridges containing resistive sensors are also common, since unwanted reactive effects due to the capacitance or inductance of connecting leads can be cancelled out.

The simplest form of AC excited impedance bridge is obtained by replacing the DC excitation voltage E in figure 3.1 by an oscillator as shown in figure 3.9. However, it is more usual to replace one side of such a bridge by a transformer, as shown in figure 3.10, since this ensures that the two sides of the bridge are excited in opposition with equal magnitude. The transformer either has a centre-tapped secondary or a pair of identical secondary windings. If the centre tap is connected to ground and the resistances are sensors a single-ended output is obtained.

A push–pull sensor is a device arranged in two parts, such that a change in the sensed parameter produces a decrease in the output of one part and

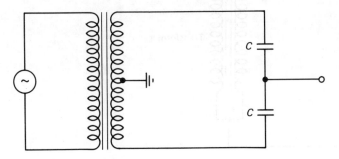

Figure 3.10 AC bridge (capacitive sensors in this example) excited via a transformer.

54 INSTRUMENTATION FOR ENGINEERS

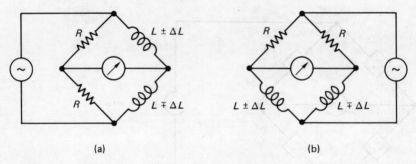

Figure 3.11 Alternative connections for push–pull sensors: (a) sensors on one side of the bridge only; (b) sensors placed both sides of detector.

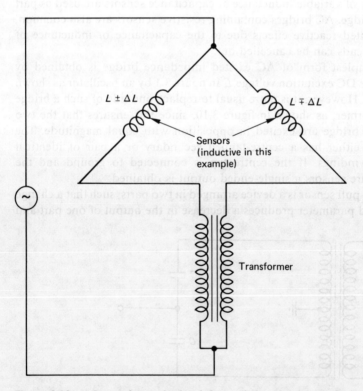

Figure 3.12 AC transformer bridge (often called a Blumlein bridge).

a simultaneous increase in the output of the second part. A number of sensors (for example, LVDTs) are available in push–pull form. A push–pull sensor can be connected into a bridge containing two resistors in one of two ways, as shown in figure 3.11. At balance an AC bridge must simultaneously satisfy two conditions. First, the real part of the out-of-balance voltage must be zero, and second, the imaginary part must be zero. If, for example, inductive sensors with a low resistance are used, the contribution to the out-of-balance voltage from changes in transducer resistance is negligible in comparison to the inductance change, and to a first approximation the bridge shown in figure 3.11(a) will give an output

$$V_{out} = (V_s/2)(\Delta L/L) \tag{3.12}$$

where V_{out} and V_s are complex (AC) voltages. Under similar conditions and if $\omega L = R$, the output of the circuit shown in figure 3.11(b) will be twice that of figure 3.11(a).

An alternative form of AC bridge which is often used with push–pull inductive sensors is shown in figure 3.12. Two of the bridge arms are inductively coupled (generally by means of a transformer), and the push–pull sensor makes up the other two arms. This circuit is known as a Blumlein bridge, and it simplifies earthing and shielding problems since stray cable capacitances appear in parallel across the measurement arms and are consequently ignored. In general, a Blumlein bridge with two closely coupled inductive arms gives better noise immunity and a greater constancy of sensitivity than an AC bridge with uncoupled inductive or purely resistive arms.

OPERATIONAL AMPLIFIER SIGNAL CONDITIONING CIRCUITS

The basic instrumentation amplifier or operational amplifier is represented by a triangle as shown in figure 3.13. It is a DC coupled differential input

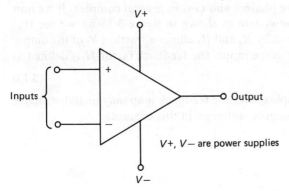

Figure 3.13 *The operational amplifier circuit symbol.*

device with a single output. This rather incomprehensible statement means essentially that it amplifies the voltage difference between the two inputs. The ideal op-amp has infinite gain, infinite input impedance, an infinite bandwidth and zero output impedance. Real op-amps approach this ideal with varying degrees of success, usually in strict proportion to the amount of money the user is willing to spend! The DC gain of an op-amp is usually in the range 10^5–10^6, and the gain reduces at 6 dB/octave as frequency increases (see the discussion of gain–bandwidth products below). The output signal is allowed to swing through most of the supply voltage range. Generally the supply voltage is split, and ± 15 V is common which allows an output swing of about 28 or 29 V.

The $+$ and $-$ symbols on figure 3.13 indicate the non-inverting ($+$) and inverting ($-$) inputs to the amplifier. When the non-inverting ($+$) input is made more positive than the inverting ($-$) input, the output becomes positive and vice versa. The $+$ and $-$ symbols do not imply that one input must be kept more positive than the other, but tell you what phase the output will have with respect to the input. For example, if a signal is connected to the negative input of an op-amp, the output will be out of phase by 180° (π radians) with respect to the input.

The most common op-amp is the μA741C, or 741 for short. It comes in a variety of packages, but the most usual form is an 8-pin DIL integrated circuit. The package and its connections are shown in figure 3.14, and a datasheet is given at the end of this chapter. The 741 is popular because (like the ideal research student) it is simple to use, tolerates maltreatment without permanent damage, performs well and is cheap to obtain.

We saw earlier that the op-amp has very high open loop (that is, without feedback) gain at 0 Hz (DC). This is done deliberately, since with a high gain amplifier the introduction of negative feedback gives an overall gain which is independent of the open loop gain. The gain in such a case is solely a function of the components used for feedback, as shown below.

Consider the ideal amplifier shown in figure 3.15(a). The gain $G = V_2/V_1$ where V_1 and V_2 are phasors and G is in general complex. If we now add some feedback to the system as shown in figure 3.15(b), we see that the potential divider formed by R_1 and R_2 allows a fraction V_f of the output voltage V_2 to be fed back to the input. The feedback factor H is defined as

$$H = V_f/V_2 \tag{3.13}$$

In general, H is complex since the feedback loop may include reactive as well as resistive components, although in this example

$$H = R_1/(R_1 + R_2)$$

which is real.

The input V_1 is now

$$V_1 = V_s + V_f$$

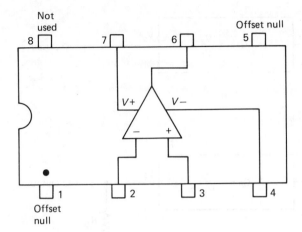
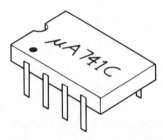

Figure 3.14 The 741 package and connections

and so the output becomes

$$V_2 = GV_1 = G(V_s + V_f) = G(V_s + HV_2)$$

that is

$$V_2 = GV_s + GHV_2 \quad (3.14)$$

So the gain of the whole system including the feedback components is A, where

$$A = V_2/V_s = G/(1 - GH) \quad (3.15)$$

If G is large (>1000) we can write

$$A = -G/GH = -1/H \quad (3.16)$$

Equation (3.16) implies that as long as the open-loop gain of an op-amp is sufficiently large, the gain of a closed loop circuit will be independent of the open-loop gain.

Some Op-Amp Terms

Before considering any op-amp circuits, it is necessary to become familiar with some of the specialised terminology associated with these devices. The

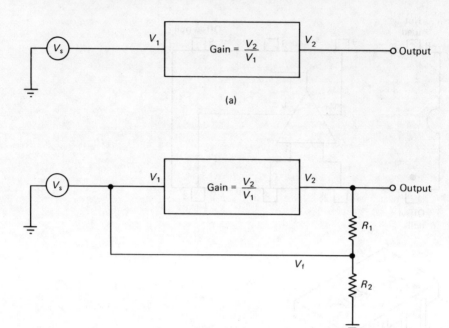

Figure 3.15 (a) Ideal amplifier without feedback. (b) Ideal amplifier with feedback.

aim of this section is to provide a brief explanation of some of the expressions used.

Input impedance: The internal impedance between the inverting and non-inverting terminals. Typical value 1 MΩ.

Output impedance: The internal impedance between the output and ground. Typical value somewhere between 10 Ω and 500 Ω.

Input-output impedance: The impedance between either input and the output is very high, and can be assumed to be infinite. An op-amp can be represented by the equivalent circuit of figure 3.16. Note that as current is drawn from the device the output voltage is reduced from its initial value V_0 and becomes

$$V_{out} = V_0 - I_L Z_2 \qquad (3.17)$$

Common Mode Gain (CMG):   An ideal op-amp amplifies only differential signals, and has a gain of zero for common-mode signals (exactly identical signals which are applied to both inputs). Real amplifiers have non-zero values of CMG. The size of the CMG is a function of the magnitudes of the common and the differential input signals. A typical CMG value for a 741 op-amp with open-loop differential gain of 100 000 is about 3.

SIGNAL CONDITIONING   59

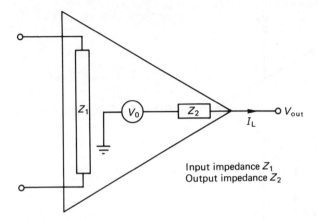

Figure 3.16   *Op-amp equivalent circuit.*

Common Mode Rejection Ratio (CMRR): CMRR is defined as the absolute value of the ratio of differential gain to common-mode gain, that is

$$\text{CMRR} = \left|\frac{A(f)}{\text{CMG}}\right| \qquad (3.18)$$

Voltage Supply Rejection Ratio (VSRR): The characteristics of an op-amp are quoted for a standard supply voltage. If this voltage changes, in general the output of the amplifier will change. The size of this output variation is expressed in terms of the equivalent differential input voltage arising from a 1 V change in supply. In other words, the input to the amplifier which would produce the same output change as that caused by a 1 V supply change is quoted as a measure of the device's sensitivity to supply variations. The VSRR for a 741 amplifier is around 20 μV/V or 94 dB.

Input Offset Voltage (IOV): For an ideal op-amp, V_{out} is zero when V_{in} is zero. Because components in symmetrical positions inside the amplifier are impossible to match precisely, this rarely occurs in practice. A pair of terminals is provided as shown in figure 3.17 for balancing out the unwanted input offset. It should be borne in mind, however, that IOV is temperature-sensitive, and balancing must therefore be carried out at the circuit's operating temperature.

Input bias current: It is necessary to supply small currents to both terminals of an op-amp to bias the transistors correctly within the amplifier. Earthing the unused input leads to an imbalance in these currents, and this results in an offset voltage appearing at the output of the device. To avoid this effect the unused terminal should be earthed through a resistor equal in value to the parallel combination of the input and feedback resistors. Figure

60 INSTRUMENTATION FOR ENGINEERS

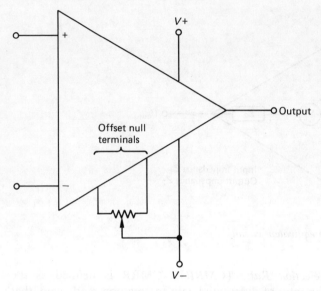

Figure 3.17 Inputs for removing unwanted offset voltages.

3.18 shows an inverting example. Biasing resistors are usually omitted from op-amp circuit diagrams, but this does not mean they are unnecessary!

Gain-bandwidth product: The gain-bandwidth product of an op-amp is the product of the DC gain and the frequency at which the gain is reduced by 3 dB, as shown in figure 3.19. Except for special-purpose devices, the roll-off of an op-amp is 6 dB/octave. If this is so, the open-loop gain-frequency diagram can also be used to calculate a closed-loop gain-bandwidth product as shown on the diagram.

Slew rate: The slew rate of an op-amp is the maximum rate of change of

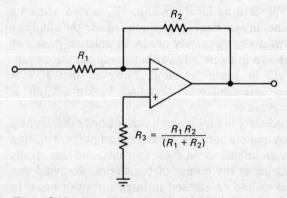

Figure 3.18 Unused input earthed through biasing resistor.

SIGNAL CONDITIONING 61

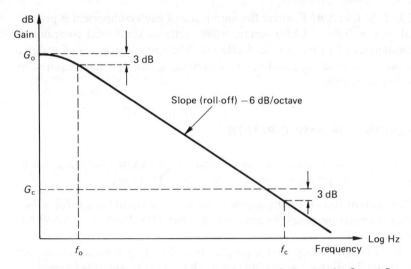

G_o = open loop gain, G_c = closed loop gain. f_o and f_c are -3 dB points for G_o and G_c respectively. Typical gain–bandwidth product for a 741 is around 10^5. Gain–bandwidth product $G_o f_o = G_c f_c$

Figure 3.19   Gain–bandwidth diagram.

output voltage with time. Suppose the output is a sine wave:

$V_{out} = V_0 \sin \omega t$

(where angular frequency $\omega = 2\pi \times$ spatial frequency f). The rate of change of output will therefore be

$$\frac{dV_{out}}{dt} = \omega V_0 \cos \omega t$$

The maximum rate of change of output ($\cos \omega t = 1$) is therefore

$$\left(\frac{dV_{out}}{dt}\right)_{max} = \omega V_0 = 2\pi f V_0$$

Thus, to give an undistorted sinusoidal output, an op-amp must be used which satisfies the condition

$\text{slew rate} \geqslant 2\pi f V_0$ volts/second   (3.19)

A typical slew rate is 0.5 V/μs (for a 741), but with special op-amps such as the NE5539 it can be as high as 800 V/μs.

Care should be taken in applying the above analysis to non-sinusoidal periodic signals. To take a common example, suppose an op-amp is used to amplify a square wave of period T. It can be shown by Fourier analysis that a square wave consists of a sum of sinusoidal components at frequencies

$1/T, 3/T, 5/T, \ldots, n/T$, where the amplitude of each component is proportional to $1/n$. Thus a 1 kHz square wave contains sinusoidal components at frequencies of 1 kHz, 3 kHz, 5 kHz etc. The spectral content of a signal must therefore be considered before determining the slew rate required by a particular circuit.

ANALYSING OP-AMP CIRCUITS

The behaviour of almost all op-amp circuits with external feedback can be understood by applying a pair of simple rules. These are:

I. The current drawn by the inputs is so small (around 0.1 μA for a 741) that for most purposes we may assume that THE INPUTS DRAW NO CURRENT.
II. Since the open-loop voltage gain is so high, a tiny voltage is sufficient to swing the output over its full range. Thus, it is a reasonable approximation to say that THE INPUTS ARE AT THE SAME VOLTAGE.

Rule II needs some further explanation. It does not mean that an op-amp actually changes the voltage at one input in response to a change at the other. (If it did, it would break rule I!). What it does mean is that the device adjusts its output so that the external feedback network brings the differential input voltage to zero (if possible).

The Inverting Amplifier

In an inverting amplifier the input signal is applied to the inverting $(-)$ input as shown in figure 3.20, and the output is out of phase by 180° (π radians) with respect to the input.

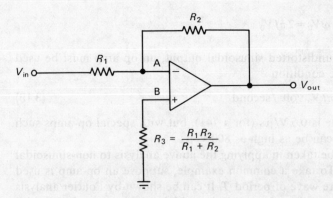

Figure 3.20   The inverting amplifier.

The op-amp rules above can be used to analyse the behaviour of an inverting amplifier:

1. Point B (see figure 3.20) is at ground, so rule II implies that A is at ground also. A point such as A is often called a virtual earth.
2. Rule I implies that the current through R_1 is equal in magnitude and has opposite sign to the current through R_2. As point A is at ground potential we can write

$$\frac{V_{out}}{R_2} = -\frac{V_{in}}{R_1}$$

3. Therefore the gain of the circuit is

$$\frac{V_{out}}{V_{in}} = \frac{-R_2}{R_1} \qquad (3.20)$$

Note that this analysis is not affected by the presence of the bias resistor R_3, because according to rule I no current flows through R_3 and hence B is at ground potential. Note also that since point A is at ground potential, the input impedance of an inverting amplifier is governed by R_1.

The Non-inverting Amplifier

The circuit of a non-inverting amplifier is shown in figure 3.21(a). In this case the signal is applied to the non-inverting (+) input, and the output is in phase with the input. Application of the rules gives

$$\text{gain} = \frac{V_{out}}{V_{in}} = 1 + \frac{R_2}{R_1} \qquad (3.21)$$

Note that if $R_2 = 0$ and R_1 is infinite, a unity-gain voltage follower results as shown in figure 3.21(b). Rule II implies that no current is drawn by the input of a non-inverting amplifier, and so no load is applied to the circuit or device providing V_{in}. A non-inverting amplifier consequently

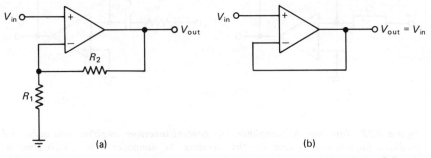

Figure 3.21 Non-inverting amplifiers: (a) non-inverting amplifier; (b) voltage follower.

64 INSTRUMENTATION FOR ENGINEERS

protects a signal source from any loads applied at V_{out}, because of the very high input impedance of the op-amp. This should be contrasted with the behaviour of the inverting amplifier, where current is drawn from V_{in} at the input and flows through resistors R_1 and R_2.

It should also be noted that a non-inverting amplifier may not be used as an attenuator—that is, it cannot have a gain of less than unity. An inverting amplifier may have a gain which is less than one.

AC Amplifiers

If only AC signals are to be amplified it is good practice to 'roll off' the gain to unity at frequencies close to DC. This prevents the amplifier becoming saturated by DC offset voltages.

The circuit for an inverting AC amplifier is shown in figure 3.22. The gain of the amplifier is $V_0/V_{in} = -Z_2/Z_1$ where Z_1 and Z_2 are impedances. From the circuit diagram:

$$Z_1 = R_1 + 1/(2\pi j f C_1) \text{ and } Z_2 = R_2 \qquad (3.22)$$

so the circuit transfer function is

$$\frac{V_{out}}{V_{in}} = \frac{-2\pi j f R_2 C}{1 + 2\pi j f R_1 C} \qquad (3.23)$$

It can be seen that for $f = 0$, $(V_{out}/V_{in}) = 0$. As $f \to \infty$, $(V_{out}/V_{in}) \to -R_2/R_1$. The -3 dB point is at $1/(2\pi R_1 C)$ Hz. For frequencies where $f > f_{3\,dB}$ the capacitor is essentially a short circuit, and the input and output impedances are the same as those of a DC coupled inverting amplifier.

Figure 3.23 shows a non-inverting AC amplifier. The transfer function in this case is

$$\frac{V_{out}}{V_{in}} = \left(\frac{1}{1 + 1/(2\pi j f R_1 C)}\right)\left(1 + \frac{R_3}{R_2}\right) \qquad (3.24)$$

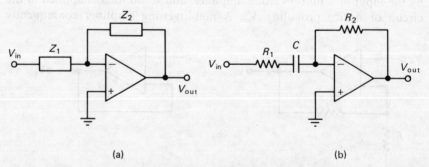

(a)   (b)

Figure 3.22 Inverting AC amplifier: (a) general inverting amplifier with input and feedback impedances Z_1 and Z_2; (b) inverting AC amplifier with -3 dB point at $f_{3\,dB} = 1/(2\pi R_1 C)$ Hz.

SIGNAL CONDITIONING 65

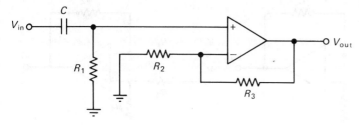

Figure 3.23 Non-inverting AC amplifier.

At high frequencies [that is, when $f > 1/(2\pi R_1 C_1)$] the gain becomes $1 + R_2/R_1$, which is similar to that of a DC coupled non-inverting amplifier.

The Constant Current Source

An op-amp current source such as that shown in figure 3.24 gives near ideal behaviour without the voltage offset inherent in transistor sources. The current I through the load is V_{in}/R, regardless of changes in load impedance.

The Differential Amplifier

Figure 3.25(a) shows a differential amplifier with equal sets of resistors for both input and feedback. The most useful property of a differential amplifier is that it rejects common-mode signals originating at the signal source. The output signal from circuit (a) is the amplified difference between the two input signals:

$$V_0 = -\frac{R_2}{R_1}(V_{s1} - V_{s2})$$

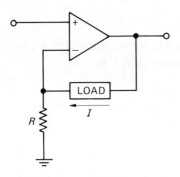

Figure 3.24 Constant current source.

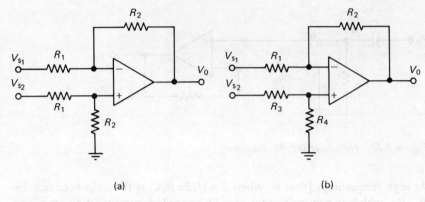

Figure 3.25 Differential amplifiers.

In circuit (b) the input and feedback resistors can have different values. The resulting output is a weighted function of the input signals:

$$V_{out} = \frac{R_4}{R_3}\left(\frac{1+R_2/R_1}{1+R_4/R_3}\right)\cdot V_{s2} - \left(\frac{R_1}{R_2}\right)\cdot V_{s1} \qquad (3.25)$$

When $R_1 = R_3$ and $R_2 = R_4$ the above expression reduces to the gain equation for circuit (a).

The differential amplifier rejects signals appearing at its input terminals if the signals are common to both inputs, that is if they have a common mode. The output in the presence of a common-mode voltage at the input terminals of the op-amp is

$$V_{out} = -\left\{\frac{R_2}{R_1}\cdot V_s + \left(1+\frac{R_2}{R_1}\right)\left(\frac{V_{COMMON}}{CMRR}\right)\right\}$$

where $V_s = V_{s1} - V_{s2}$. For a common-mode voltage at the signal source, the output voltage is

$$V_{out} = -\frac{R_2}{R_1}\left(V_s + \frac{V_{COMMON}}{CMRR}\right)$$

To maximise the rejection of any common mode signals applied to V_{s1} and V_{s2}, component values should ideally be chosen such that

$$\frac{R_1}{R_2} = \frac{R_3}{R_4} \quad \text{and} \quad R_1 = R_3 + R_4$$

The Summing Amplifier

The output of the circuit of figure 3.26 is a sum of the applied voltages V_1,

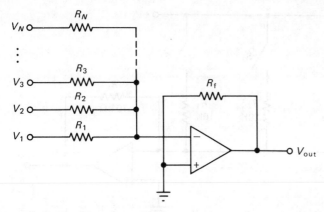

Figure 3.26 Summing amplifier.

V_2 etc. weighted by the resistor values as shown in the following expression:

$$\frac{V_{out}}{R_f} = -\left\{\frac{V_1}{R_1} + \frac{V_2}{R_2} + \cdots + \frac{V_N}{R_N}\right\} \tag{3.26}$$

If the resistors all have the same value this reduces to

$$V_{out} = -(V_1 + V_2 + \cdots + V_N) \tag{3.27}$$

The circuit may be extended to have as many inputs as required. If resistor values are chosen which give a natural binary weighting, the circuit can be used as a simple D/A converter as shown in figure 3.27. However, the wide range of resistor values needed to provide a binary weighting prevents expansion to more than about 4 bits, because of temperature tracking problems.

The Precision Rectifier

A diode used as a half-wave rectifier is of limited application in low-voltage circuits, because of the voltage drop across a diode during forward conduction. With a silicon diode at least 0.7 V has to be applied before any output voltage appears. The circuit of figure 3.28(a) avoids this problem. When the input is positive the high gain of the op-amp causes the diode to conduct when the non-inverting input is only a few microvolts more positive than the inverting input.

The circuit shown in 3.28(a) suffers from two limitations. First, during negative input excursions a large differential voltage is applied to the op-amp, driving it into saturation. An op-amp may take up to 50 μs to recover from saturation, and this places limits on the high-frequency performance of the circuit. Second, an op-amp which can tolerate large differential inputs without damage must be used. These can be expensive.

68 INSTRUMENTATION FOR ENGINEERS

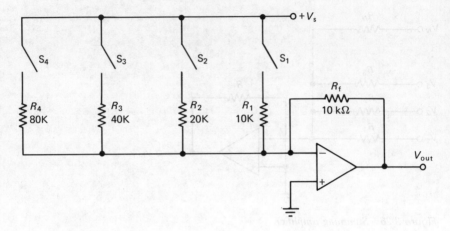

$$\frac{V_{out}}{R_f} = \frac{a_1 V_s}{R_1} + \frac{a_2 V_s}{R_2} + \frac{a_3 V_s}{R_3} + \frac{a_4 V_s}{R_4}$$

where $a_1 = 1$ for S_1 closed and $a_1 = 0$ for S_1 open and so on.
The binary input is determined by the settings of S_1-S_4

Figure 3.27 Digital-to-analogue converter based on summing amplifier.

The arrangement of figure 3.28(b) avoids these difficulties. Diode D_2 prevents the op-amp becoming saturated, and a much faster response is achieved.

The Integrator

An op-amp can be used to make an almost perfect integrator. The circuit used is shown in figure 3.29. Applying the op-amp rules, we see that the input current V_{in}/R flows through C. As the inverting input is at ground potential (rule II), the output voltage is found from

$$V_{in}/R = -C(dV_{out}/dt)$$

or

$$V_{out} = -\frac{1}{RC} \int V_{in} \, dt + \text{constant} \qquad (3.28)$$

The constant of integration is of course determined by the initial conditions, and can appear as a DC voltage added to the output signal.

An integrator is a useful circuit if a 90° phase shift is required, since the integral of a cosine function is a sine wave. This is often done to test the purity of a sinusoid—the signal and its integral are used to form a Lissajous figure on an oscilloscope, and if the gain of the integration is unity, a perfect circle results from a pure sinusoid.

SIGNAL CONDITIONING   69

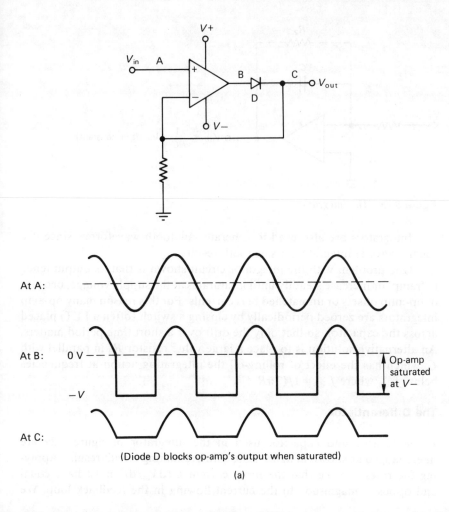

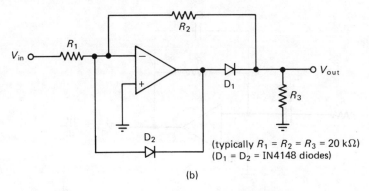

Figure 3.28 Op-amp precision rectifiers: (a) simple precision rectifier; (b) practical precision rectifier.

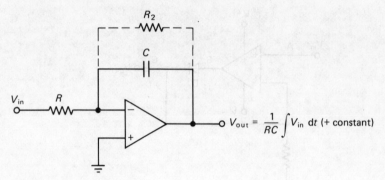

Figure 3.29 The integrator.

Integrators are also used to generate saw-tooth waveforms, since if a square wave is integrated, a saw-tooth results.

One problem with the integrator circuit shown is that its output tends to 'ramp' or increase steadily until it saturates at the supply voltage, because of op-amp offsets or unmatched bias currents. For this reason many op-amp integrators are zeroed periodically by closing a switch (often a FET) placed across the capacitor, so that only the drift over a short time period matters. An alternative solution is to place a large value resistor R_2 in parallel with C, which has the effect of rolling-off the integrating action at frequencies below $f_{3\,\mathrm{dB}}$ where $f_{3\,\mathrm{dB}} = 1/(2\pi R_2 C)$.

The Differentiator

If the resistor and capacitor used in the integrator of figure 3.29 are interchanged as shown in figure 3.30, a differentiating circuit results. Applying the rules, we see that the input current $C(\mathrm{d}V_{\mathrm{in}}/\mathrm{d}t)$ must have equal and opposite magnitude to the current flowing in the feedback loop. We

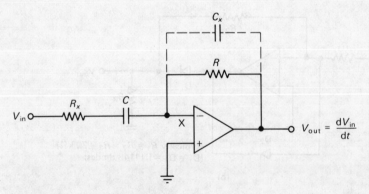

Figure 3.30 The differentiator.

also see that rule II implies point X is at ground potential (that is, a virtual earth). Hence

$$V_{out}/R = -C(dV_{in}/dt)$$

or

$$V_{out} = -RC(dV_{in}/dt) \tag{3.29}$$

Differentiators can, however, be awkward to use as they tend to amplify noise and spurious voltage spikes at the input. This is because, obviously, the differential of a rapidly changing signal such as a spike is likely to be large. In addition, differentiators can suffer from instability problems at high frequencies. Both of these shortcomings may be overcome to some extent by rolling-off the differentiating action at high frequencies. This is achieved by the addition of R_x and C_x to the circuit, as shown on figure 3.30.

Current-to-Voltage Converter

The simplest current-to-voltage converter is, of course, a resistor. However, the trouble with using resistors is that they present a non-zero impedance to the source of input current. This can cause difficulties if the current source has very little compliance. (A constant current source can only maintain a constant current through a load over a finite range of load voltage. The output voltage range over which a current source is well behaved is known as its compliance.) Photovoltaic cells, for example, have a very small compliance.

Figure 3.31 shows how to use an op-amp as a current-to-voltage converter. The inverting input is a virtual earth. The output voltage is determined by the feedback resistor; in the circuit of figure 3.31 it is 1 V per μA of input current. The resistor between the non-inverting input and ground could probably be omitted, but is desirable to match the input bias currents as explained earlier.

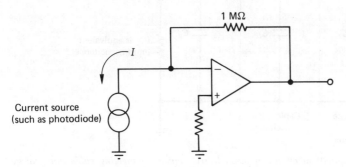

Figure 3.31 Current-to-voltage converter.

The Charge Amplifier

A charge amplifier is usually used with capacitive sensors, and for piezoelectric sensors which act as a charge source. A charge amplifier is only sensitive to variations in charge, which means that almost any length of cable can be used to connect a sensor to a charge amplifier without affecting the sensitivity. As shown in figure 3.32, a charge amplifier consists of an op-amp with capacitive feedback. This gives the circuit an input capacitance $C_P = C_F(G-1)$, where G is the open loop gain of the op-amp. The amplifier output voltage is given by

$$V_{out} = \frac{QG}{C_S + C_C - C_F(G-1)} \tag{3.30}$$

Since the open loop gain of an op-amp is very high the output voltage becomes

$$V_{out} = Q/C_F \tag{3.31}$$

With C_F constant the output voltage V_{out} is directly proportional to the input charge Q. The cable capacitance C_C normally has a negligible effect. Only when the cable is so long that the size of C_C approaches C_F will the sensitivity of the circuit be affected. The feedback resistor R_F is there to provide a suitable input bias current, and should be chosen so that $R_F \geq 1/(2\pi f C_F)$ where f is the operating frequency.

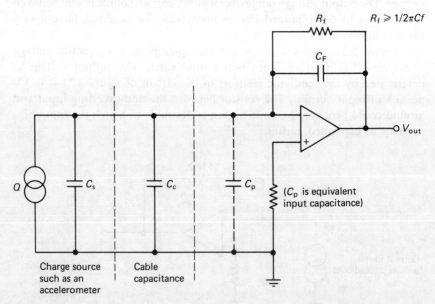

Figure 3.32 Equivalent circuit of piezoelectric sensor, connecting cable and charge amplifier.

The Comparator

A comparator is simply an open-loop op-amp, which is driven into positive or negative saturation according to the difference between the two voltages at its inputs. Since the open loop gain of an op-amp is so high (typically 10^5 for a 741), the input voltages have to be equal to within a small fraction of a millivolt for the device not to be saturated. The polarity of the saturated output indicates the direction of the inequality relating the input voltages. Although ordinary op-amps can be used as comparators, a device such as the 741 takes an appreciable time to recover from saturation. Its use as a high-speed comparator will also be limited by the slew rate (typically 10^6 V/second for a 741). It is much better to use a special op-amp which is designed to act as a comparator. Examples include the LM306, LM311 and LM393 produced by National Semiconductor. These have a very fast recovery time and slew rates in excess of 10^9 V/second.

There are three points to note about comparators. Since there is no feedback, rule II is not obeyed. The inputs are not necessarily at the same voltage. Again because there is no feedback, the input impedance is not necessarily constant. As a result the input signal sees a changing load and a changing (small) input current as the comparator output switches. Finally, some comparators will only permit limited differential inputs, as small as 5 V in some cases. Check the device datasheet before deciding to use it in a design!

LINEAR INTEGRATED CIRCUITS
TYPES SN52741, SN72741 GENERAL-PURPOSE OPERATIONAL AMPLIFIERS

BULLETIN NO. DL-S 7311363, NOVEMBER 1970—REVISED SEPTEMBER 1973

- Short-Circuit Protection
- Offset-Voltage Null Capability
- Large Common-Mode and Differential Voltage Ranges
- No Frequency Compensation Required
- Low Power Consumption
- No Latch-up
- Same Pin Assignments as SN52709/SN72709

description

The SN52741 and SN72741 are general-purpose operational amplifiers, featuring offset-voltage null capability.

The high common-mode input voltage range and the absence of latch-up make the amplifier ideal for voltage-follower applications. The devices are short-circuited protected and the internal frequency compensation ensures stability without external components. A low-value potentiometer may be connected between the offset null inputs to null out the offset voltage as shown in Figure 11.

The SN52741 is characterized for operation over the full military temperature range of –55°C to 125°C; the SN72741 is characterized for operation from 0°C to 70°C.

schematic

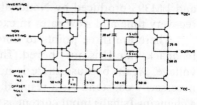

COMPONENT VALUES SHOWN ARE NOMINAL

terminal assignments

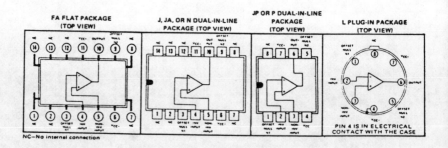

NC–No internal connection

TEXAS INSTRUMENTS

SIGNAL CONDITIONING

TYPES SN52741, SN72741
GENERAL-PURPOSE OPERATIONAL AMPLIFIERS

absolute maximum ratings over operating free-air temperature range (unless otherwise noted)

		SN52741	SN72741	UNIT
Supply voltage V_{CC+} (see Note 1)		22	18	V
Supply voltage V_{CC-} (see Note 1)		−22	−18	V
Differential input voltage (see Note 2)		±30	±30	V
Input voltage (either input, see Notes 1 and 3)		±15	±15	V
Voltage between either offset null terminal (N1/N2) and V_{CC-}		±0.5	±0.5	V
Duration of output short-circuit (see Note 4)		unlimited	unlimited	
Continuous total power dissipation at (or below) 25°C free-air temperature (see Note 5)		500	500	mW
Operating free-air temperature range		−55 to 125	0 to 70	°C
Storage temperature range		−65 to 150	−65 to 150	°C
Lead temperature 1/16 inch from case for 60 seconds	FA, J, JA, JP, or L package	300	300	°C
Lead temperature 1/16 inch from case for 10 seconds	N or P package	260	260	°C

NOTES: 1. All voltage values, unless otherwise noted, are with respect to the zero reference level (ground) of the supply voltages where the zero reference level is the midpoint between V_{CC+} and V_{CC-}.
2. Differential voltages are at the noninverting input terminal with respect to the inverting input terminal.
3. The magnitude of the input voltage must never exceed the magnitude of the supply voltage or 15 volts, whichever is less.
4. The output may be shorted to ground or either power supply. For the SN52741 only, the unlimited duration of the short-circuit applies at (or below) 125°C case temperature or 75°C free-air temperature.
5. For operation above 25°C free-air temperature, refer to Dissipation Derating Curve, Figure 12.

electrical characteristics at specified free-air temperature, $V_{CC+} = 15$ V, $V_{CC-} = -15$ V

PARAMETER		TEST CONDITIONS†		SN52741			SN72741			UNIT
				MIN	TYP	MAX	MIN	TYP	MAX	
V_{IO}	Input offset voltage	$R_S \leq 10$ kΩ	25°C		1	5		1	6	mV
			Full range			6			7.5	
$\Delta V_{IO(adj)}$	Offset voltage adjust range		25°C		±15			±15		mV
I_{IO}	Input offset current		25°C		20	200		20	200	nA
			Full range			500			300	
I_{IB}	Input bias current		25°C		80	500		80	500	nA
			Full range			1500			800	
V_I	Input voltage range		25°C	±12	±13		±12	±13		V
			Full range	±12			±12			
V_{OPP}	Maximum peak-to-peak output voltage swing	$R_L = 10$ kΩ	25°C	24	28		24	28		V
		$R_L \geq 10$ kΩ	Full range	24			24			
		$R_L = 2$ kΩ	25°C	20	26		20	26		
		$R_L \geq 2$ kΩ	Full range	20			20			
A_{VD}	Large-signal differential voltage amplification	$R_L \geq 2$ kΩ, $V_O = \pm10$ V	25°C	50,000	200,000		20,000	200,000		
			Full range	25,000			15,000			
r_i	Input resistance		25°C	0.3	2		0.3	2		MΩ
r_o	Output resistance	$V_O = 0$ V, See Note 6	25°C		75			75		Ω
C_i	Input capacitance		25°C		1.4			1.4		pF
CMRR	Common-mode rejection ratio	$R_S \leq 10$ kΩ	25°C	70	90		70	90		dB
			Full range	70			70			
$\Delta V_{IO}/\Delta V_{CC}$	Supply voltage sensitivity	$R_S \leq 10$ kΩ	25°C		30	150		30	150	µV/V
			Full range			150			150	
I_{OS}	Short-circuit output current		25°C		±25	±40		±25	±40	mA
I_{CC}	Supply current	No load, No signal	25°C		1.7	2.8		1.7	2.8	mA
			Full range			3.3			3.3	
P_D	Total power dissipation	No load, No signal	25°C		50	85		50	85	mW
			Full range			100			100	

†All characteristics are specified under open-loop operation. Full range for SN52741 is −55°C to 125°C and for SN72741 is 0°C to 70°C.
NOTE 6: This typical value applies only at frequencies above a few hundred hertz because of the effects of drift and thermal feedback.

TEXAS INSTRUMENTS

TYPES SN52741, SN72741
GENERAL-PURPOSE OPERATIONAL AMPLIFIERS

operating characteristics, $V_{CC+} = 15$ V, $V_{CC-} = -15$ V, $T_A = 25°C$

PARAMETER		TEST CONDITIONS	SN52741			SN72741			UNIT
			MIN	TYP	MAX	MIN	TYP	MAX	
t_r	Rise time	$V_I = 20$ mV, $R_L = 2$ kΩ, $C_L = 100$ pF, See Figure 1		0.3			0.3		μs
	Overshoot			5%			5%		
SR	Slew rate at unity gain	$V_I = 10$ V, $R_L = 2$ kΩ, $C_L = 100$ pF, See Figure 1		0.5			0.5		V/μs

DEFINITION OF TERMS

Input Offset Voltage (V_{IO}) The d-c voltage which must be applied between the input terminals to force the quiescent d-c output voltage to zero. The input offset voltage may also be defined for the case where two equal resistances (R_S) are inserted in series with the input leads.

Input Offset Current (I_{IO}) The difference between the currents into the two input terminals with the output at zero volts.

Input Bias Current (I_{IB}) The average of the currents into the two input terminals with the output at zero volts.

Input Voltage Range (V_I) The range of voltage that if exceeded at either input terminal will cause the amplifier to cease functioning properly.

Maximum Peak-to-Peak Output Voltage Swing (V_{OPP}) The maximum peak-to-peak output voltage which can be obtained without waveform clipping when the quiescent d-c output voltage is zero.

Large-Signal Differential Voltage Amplification (A_{VD}) The ratio of the peak-to-peak output voltage swing to the change in differential input voltage required to drive the output.

Input Resistance (r_i) The resistance between the input terminals with either input grounded.

Output Resistance (r_o) The resistance between the output terminal and ground.

Input Capacitance (C_i) The capacitance between the input terminals with either input grounded.

Common-Mode Rejection Ratio (CMRR) The ratio of differential voltage amplification to common-mode voltage amplification. This is measured by determining the ratio of a change in input common-mode voltage to the resulting change in input offset voltage.

Supply Voltage Sensitivity ($\Delta V_{IO}/\Delta V_{CC}$) The ratio of the change in input offset voltage to the change in supply voltages producing it. For these devices, both supply voltages are varied symmetrically.

Short-Circuit Output Current (I_{OS}) The maximum output current available from the amplifier with the output shorted to ground or to either supply.

Total Power Dissipation (P_D) The total d-c power supplied to the device less any power delivered from the device to a load. At no load: $P_D = V_{CC+} \cdot I_{CC+} + V_{CC-} \cdot I_{CC-}$.

Rise Time (t_r) The time required for an output voltage step to change from 10% to 90% of its final value.

Overshoot The quotient of: (1) the largest deviation of the output signal value from its steady-state value after a step-function change of the input signal, and (2) the difference between the output signal values in the steady state before and after the step-function change of the input signal.

Slew Rate (SR) The average time rate of change of the closed-loop amplifier output voltage for a step-signal input. Slew rate is measured between specified output levels (0 and 10 volts for this device) with feedback adjusted for unity gain.

TEXAS INSTRUMENTS

SIGNAL CONDITIONING 77

TYPES SN52741, SN72741
GENERAL-PURPOSE OPERATIONAL AMPLIFIERS

PARAMETER MEASUREMENT INFORMATION

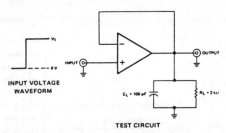

INPUT VOLTAGE WAVEFORM

TEST CIRCUIT

FIGURE 1—RISE TIME, OVERSHOOT, AND SLEW RATE

TYPICAL CHARACTERISTICS

INPUT OFFSET CURRENT
vs
FREE-AIR TEMPERATURE

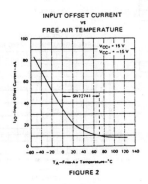

FIGURE 2

INPUT BIAS CURRENT
vs
FREE-AIR TEMPERATURE

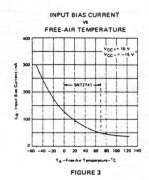

FIGURE 3

MAXIMUM PEAK-TO-PEAK OUTPUT VOLTAGE
vs
LOAD RESISTANCE

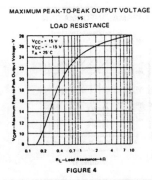

FIGURE 4

MAXIMUM PEAK-TO-PEAK OUTPUT VOLTAGE
vs
FREQUENCY

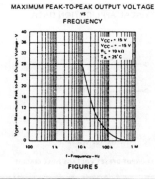

FIGURE 5

TEXAS INSTRUMENTS

TYPES SN52741, SN72741
GENERAL-PURPOSE OPERATIONAL AMPLIFIERS

TYPICAL CHARACTERISTICS

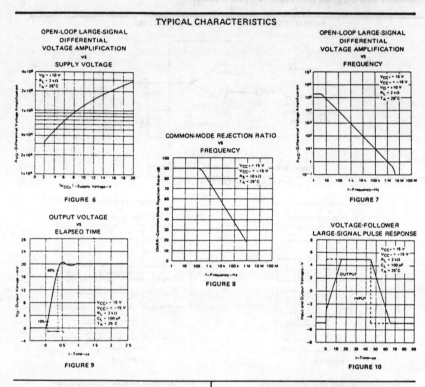

FIGURE 6 — OPEN-LOOP LARGE-SIGNAL DIFFERENTIAL VOLTAGE AMPLIFICATION vs SUPPLY VOLTAGE

FIGURE 7 — OPEN-LOOP LARGE-SIGNAL DIFFERENTIAL VOLTAGE AMPLIFICATION vs FREQUENCY

FIGURE 8 — COMMON-MODE REJECTION RATIO vs FREQUENCY

FIGURE 9 — OUTPUT VOLTAGE vs ELAPSED TIME

FIGURE 10 — VOLTAGE-FOLLOWER LARGE-SIGNAL PULSE RESPONSE

TYPICAL APPLICATION DATA

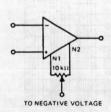

FIGURE 11 — INPUT OFFSET VOLTAGE NULL CIRCUIT

THERMAL INFORMATION

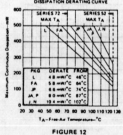

FIGURE 12 — DISSIPATION DERATING CURVE

TEXAS INSTRUMENTS

Chapter 4
Analogue Filters

INTRODUCTION

The term filtering has the commonly accepted meaning of separation—something is retained and something is rejected from the record being filtered. In electronic instrumentation signals are often filtered to improve the signal-to-noise ratio. As we have seen, signals are usually voltages, and in chapters 7 and 8 it will be shown that a signal can be made up from a large number of different frequency components. Usually only some of these frequency components are wanted. The others (which may interfere with the measurement) are removed by filtering.

The unwanted frequency components in a signal are rejected by passing the signal through a circuit which is designed to attenuate certain bandwidths while not affecting others. Such a circuit is called a filter, or more properly an analogue filter to distinguish it from the digital filtering process carried out by a computer-based system.

Put another way, the purpose of an analogue filter is to allow signals in certain frequency ranges to be transmitted through a system while preventing the transmission of signals in other frequency ranges. There are four fundamental types of filter as shown in figure 4.1, which illustrates the characteristics of perfect low-pass, high-pass, bandpass and bandstop filters. Ideally a signal is not attenuated at all in the pass band, and is completely attenuated in the stop band. The phase shift within the passband of an ideal filter is a linear function of frequency.

In general, a bandpass filter may be constructed by connecting a low-pass and a high-pass filter with appropriate cut-off frequencies in series as shown by figure 4.2. A bandstop filter can be made by connecting a low-pass and a high-pass filter in parallel, again with appropriate cut-off

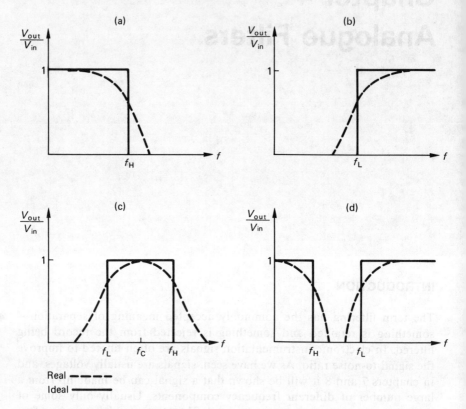

Figure 4.1 Ideal and real filter transfer functions (- - - - real, ——— ideal): (a) low-pass; (b) high-pass; (c) bandpass; (d) bandstop.

frequencies. However, for lower-order filters the high-pass and low-pass filter combinations needed are often constructed around the same op-amp(s).

Perfect filters can never be built, and real filters always have non-ideal characteristics. Typical real filter transfer functions are shown in figure 4.1. It is common to find that signals are not completely attenuated in the stop band, and non-linear phase changes can occur. However, sufficiently good approximations to ideal behaviour can be achieved to make the construction of active filter circuits one of the most important applications of op-amps. The 'corner frequency' of a real filter is defined as the point at which the amplitude of the response has fallen by 3 dB compared with the amplitude at the centre of the passband. For this reason the corner frequency is often known as the −3 dB point.

Filter design is a specialised subject, and it should be noted that much more sophisticated filters can be built than those described in this chapter. References to specialist books on filter design will be found in the bibliogra-

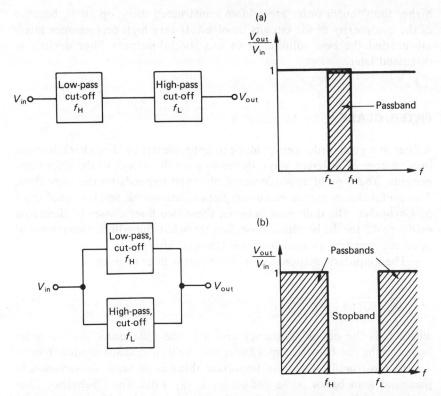

Figure 4.2 Bandpass and bandstop filters using high-pass and low-pass circuits: (a) bandpass filter synthesis; (b) bandstop filter synthesis.

phy. However, the designs outlined here should suffice for most practical applications.

FILTER ORDER

The simplest kind of active filter contains one capacitor for each cut-off frequency—that is, one capacitor for low and high pass, and two for bandpass and bandstop filters. This type is called a first-order filter. The next degree of complexity is obtained by a second-order filter (which contains two capacitors in its high-pass and low-pass forms, and four for a bandpass system), and so on. The higher the order of the filter, the closer it approaches ideal characteristics, in that the roll-off or attenuation outside the passband is steeper. As a rule of thumb the attenuation outside the passband of a filter in decibels (dB) per octave will be six times the filter order. Thus, a third-order filter will have a roll-off of 18 dB/octave. Filters higher than fourth order are seldom constructed using op-amps, because of the complexity of the circuits involved. If very high performance filters are needed the best solution is to use special-purpose filter devices as discussed later.

FILTER CLASS

A filter of a given order can be made to approximate to ideal characteristics in a number of different ways, depending on the values of the filter components. The type of approximation obtained depends on the filter class. Two useful classes are the *maximally flat* or *Butterworth*, and the *equal-ripple* or *Chebyshev*. The difference between these two filter classes is illustrated on figure 4.3 for the bandpass type. Similar behaviour will be superimposed upon the transfer functions of other types of filter.

The amplitude response of a Butterworth filter is given by

$$\frac{V_{\text{out}}}{V_{\text{in}}} = \frac{1}{\sqrt{(1+(f/f_{\text{C}})^{2n})}} \quad (4.1)$$

where f_c is the cut-off frequency and n is the filter order. As the order increases the passband becomes flatter and the stopband attenuation steeper.

In most applications the important thing is to keep the variation in passband gain below some known level, say 1 dB. The Chebyshev filter class responds to this reality by allowing the passband gain to vary in 'ripples' as shown in figure 4.3. By allowing some passband ripple, a much sharper cut-off is obtained. The amplitude response of a Chebyshev filter is

$$\frac{V_{\text{out}}}{V_{\text{in}}} = \frac{1}{\sqrt{(1+E^2 C_n^2(f/f_{\text{C}}))}} \quad (4.2)$$

(where E is a constant which defines the ripple and C_n is a polynomial).

Comparing the second-order Butterworth filter with the second-order Chebyshev filter, we see that the Butterworth type gives the better approximation in the passband, but attenuates signals in the stop band less well than the Chebyshev type. The Chebyshev type however has a 'ripple' in its passband transfer function, as shown in figure 4.3. The size of the ripple is usually quoted in dB. The form of the low-pass and high-pass transfer functions for the two classes can be deduced by inspection of figure 4.3.

The order and class of a filter for a given application is usually selected in terms of the steady-state frequency response. However, it should be remembered that the transient response of a filter may be quite different from its steady-state behaviour, and frequently some experimentation is necessary when designing filters for transient applications.

ANALOGUE FILTERS 83

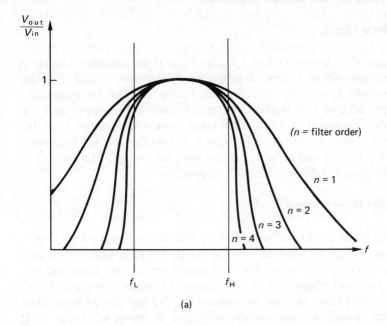

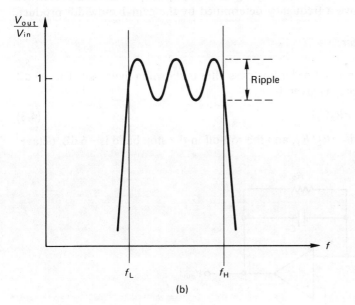

Figure 4.3 Filter classes (bandpass example): (a) Butterworth class bandpass filter; (b) Chebyshev class bandpass filter.

OPERATIONAL-AMPLIFIER FILTERS

Normalised Filters

The design of higher-order filters is simplified if an approach known as *normalised filter design* is used. A simple set of equations is used to obtain the normalised filter, which is defined as having a cut-off frequency of 1 radian per second (=0.16 Hz), and which uses 1 Ω resistors and 1 F capacitors. The design of a filter then takes place in two stages. First, the component values are changed to achieve the desired cut-off frequency. Second, the resistor and capacitor values used are changed in a prescribed manner to obtain realistic values of components.

First-Order Filter Design

First-order filters can be designed directly without the intermediate step of normalised design. The roll-off outside the passband is 6 dB/octave in a first-order filter. The non-ideal characteristics of any op-amp mean that bias currents and gain effects must be considered in practical circuits. A DC analysis of bias currents must be undertaken, and the unused input must be earthed through an appropriate resistor as discussed in chapter 3. It must also be remembered that the gain of any op-amp circuit using feedback decreases above a frequency determined by the gain–bandwidth product.

Low-pass filter

The circuit for a first-order low-pass filter is shown in figure 4.4. The −3 dB cut-off frequency is given by

$$f_H = 1/(2\pi R_2 C_2) \tag{4.3}$$

The DC gain is $-R_2/R_1$, and the roll-off in the stop band is −6 dB/octave.

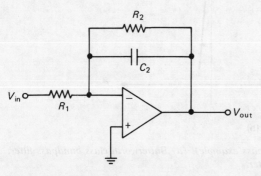

Figure 4.4 First-order low-pass filter.

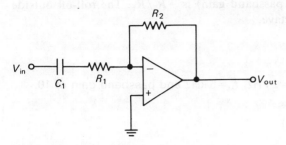

Figure 4.5 First-order high-pass filter.

Example

Design a low-pass filter with $f_H = 2$ kHz and DC gain of 5:
$$1/(2\pi R_2 C_2) = 2000$$
$$R_2/R_1 = 5$$

We have two equations with three unknowns. We can therefore choose one of the three components arbitrarily. Let $C_2 = 0.01$ μF. Then from the equations above, $R_2 = 7.96$ kΩ and $R_1 = 1.59$ kΩ.

High-pass filter

The circuit diagram of a first-order high-pass filter is shown in figure 4.5. The -3 dB cut-off frequency is

$$f_L = 1/(2\pi R_1 C_1) \tag{4.4}$$

and the high-frequency gain beyond the cut-off frequency is R_2/R_1. The roll-off below f_L is -6 dB/octave.

Bandpass filter

The circuit diagram of a first-order bandpass filter is shown in figure 4.6. The low and high cut-off frequencies are respectively

$$f_L = 1/(2\pi R_2 C_2) \quad \text{and} \quad f_H = 1/(2\pi R_1 C_1) \tag{4.5}$$

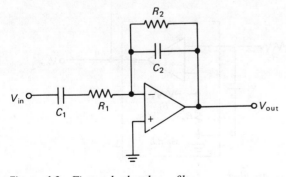

Figure 4.6 First-order bandpass filter.

The mid-frequency gain (passband gain) is $-R_2/R_1$. The roll-off outside the passband is -6 dB/octave.

Example

Design a filter with $f_L = 2$ kHz, $f_H = 5$ kHz and passband gain of 10. We have

$$2000 = 1/(2\pi R_2 C_2)$$

$$5000 = 1/2\pi R_1 C_1)$$

$$R_2/R_1 = 10$$

There are three equations and four unknowns, so we choose one of the four arbitrarily. Let $C_2 = 0.01$ μF. Then $R_2 = 7.96$ kΩ, $R_1 = 796$ Ω and $C_1 = 0.04$ μF.

Second-Order Filters

A second-order low-pass filter is shown in figure 4.7. The passband gain is $K = 1 + R_b/R_a$. The transfer function of the filter is determined by the amplifier gain and by the values of the resistors and capacitors. Three choices of K are particularly useful:

(a) $K = 2$. In this case $R_a = R_b$, which gives good matching and temperature tracking for R_a and R_b.
(b) $K = 1$. This results in a voltage follower (see chapter 3) and eliminates the need for R_a and R_b. R_a is replaced with an open circuit and R_b with a short circuit.
(c) $R_1 = R_2$ and $C_1 = C_2$. This again results in good matching. When equal-valued components are used costs are reduced, which is important if large numbers of filters are to be made.

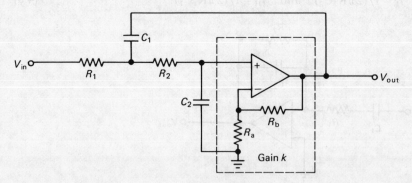

Figure 4.7   Second-order low-pass filter.

ANALOGUE FILTERS

Table 4.1

Filter class	R_1	R_2	C_1	C_2	K
Maximally flat	1.000 00	1.000 00	0.874 03	1.144 12	2.000 00
(Butterworth)	1.000 00	1.000 00	1.414 21	0.707 11	1.000 00
3.01 dB at ω_H	1.000 00	1.000 00	1.000 00	1.000 00	1.585 78
Equal ripple	1.000 00	1.000 00	0.770 88	0.855 557	2.000 00
(Chebyshev)	1.000 00	1.000 00	1.402 59	0.470 13	1.000 00
0.5 dB ripple	0.812 20	0.812 20	1.000 00	1.000 00	1.542 13
Equal ripple	1.000 00	1.000 00	0.938 09	0.966 88	2.000 00
(Chebyshev)	1.000 00	1.000 00	1.821 92	0.497 83	1.000 00
1 dB ripple	0.952 37	0.952 37	1.000 00	1.000 00	1.954 46

Second-order filter design may be carried out using the normalised filter concept mentioned earlier. The design is first carried out for the cut-off frequency $f_H = 0.159$ Hz ($\omega_H = 1$ radian/second). The component values are then changed in two steps: the first step gives the correct cut-off frequency and the second step gives reasonable values of resistors and capacitors. Table 4.1 lists gain and component values for the three choices (a), (b) and (c) above for $\omega_H = 1$, for a Butterworth (maximally flat) and two Chebyshev filters, one with 0.5 dB and one with 1 dB ripple.

Example

Design a low-pass second-order Butterworth filter with a cut-off frequency of 2500 Hz and a DC gain of 1.

The normalised filter has $R_1 = R_2 = 1\,\Omega$, $C_1 = 1.414$ F, and $C_2 = 0.707$ F, as obtained from table 4.1. To obtain a cut-off frequency of 2500 Hz $= 2\pi \times 2500$ radians/second, we divide all the capacitor values by $2\pi \times 2500$, giving

$C_1 = 90\,\mu$F and $C_2 = 45\,\mu$F

Finally, to obtain a set of reasonable component values, we multiply the resistor values by 10^5 and divide the capacitor values by 10^5 (so long as the factors are the same, the filter characteristics are unaffected). For economy we make $R_a = R_b = R_1 = R_2$. The component values are therefore:

$R_a = R_b = R_1 = R_2 = 100$ kΩ

$C_1 = 0.9$ nF

$C_2 = 0.45$ nF

In practice, the designer must use commercially available capacitors. This means that C_1 will probably be 1 nF, and $C_2 = 0.47$ nF.

Since these are not exactly equal to the calculated values, the cut-off frequency will differ slightly from the one specified.

Second-Order High-Pass Filter

Figure 4.8 shows a circuit for a second-order high-pass filter. Component values for the normalised filter are listed in table 4.2. The procedure to be followed is the same as that used for the low-pass case:

(a) Components are selected from the table for the desired filter class, for $\omega_L = 1$.
(b) The capacitor values are divided by the desired cut-off frequency in radians/second.
(c) Resistance values are multiplied and capacitance values divided by a suitable factor to give practical component values.

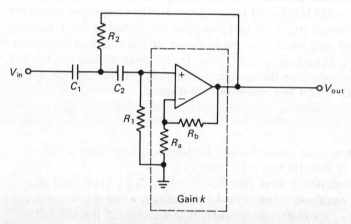

Figure 4.8 Second-order high-pass filter.

Table 4.2

Filter class	R_1	R_2	C_1	C_2	K
Maximally flat	1.000 00	1.000 00	1.414 21	0.707 11	2.000 00
(Butterworth)	0.707 11	1.414 21	1.000 00	1.000 00	1.000 00
3.01 dB at ω_1	1.000 00	1.000 00	1.000 00	1.000 00	1.585 79
Equal ripple	1.000 00	1.000 00	1.425 63	1.063 56	2.000 00
(Chebyshev)	0.712 81	2.127 07	1.000 00	1.000 00	1.000 00
0.5 dB ripple	1.231 34	1.231 34	1.000 00	1.000 00	1.842 2
Equal ripple	1.000 00	1.000 00	1.097 72	1.004 36	2.000 00
(Chebyshev)	0.545 86	2.008 72	1.000 00	1.000 00	1.000 00
1 dB ripple	1.050 01	1.050 01	1.000 00	1.000 00	1.000 00

ANALOGUE FILTERS 89

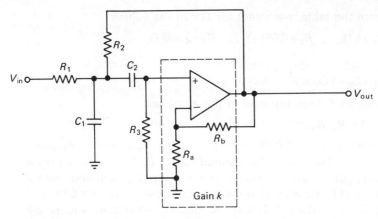

Figure 4.9 Second-order bandpass filter.

Table 4.3

Q	R_1	R_2	R_3	C_1	C_2	K
2	1.414 21	1.414 21	1.414 21	1.000 00	1.000 00	3.292 84
	1.000 00	0.740 31	2.350 78	1.000 00	1.000 00	2.000 00
5	1.414 21	1.414 21	1.414 21	1.000 00	1.000 00	3.717 16
	1.000 00	0.634 39	2.576 30	1.000 00	1.000 00	2.000 00
10	1.414 21	1.414 21	1.414 21	1.000 00	1.000 00	3.858 58
	1.000 00	0.604 71	2.635 87	1.000 00	1.000 00	2.000 00
20	1.414 21	1.414 21	1.414 21	1.000 00	1.000 00	3.924 28
	1.000 00	0.590 76	2.692 74	1.000 00	1.000 00	2.000 00

(d) Values are rounded off to those commercially available; if exact values are required, trimmers will be necessary.

Second-Order Bandpass Filter

Figure 4.9 shows the circuit of a second-order bandpass filter. A bandpass filter may be characterised by its centre frequency f_C and by the Q of the filter. The Q of a filter is defined as the ratio of the centre frequency f_C to the bandwidth $f_H - f_L$, that is

$$Q = f_C / (f_H - f_L) \tag{4.6}$$

Table 4.3 lists values of K and of components for four different values of Q for the normalised filter.

Example

Design a filter with a centre frequency of 5 kHz, a gain of 2 and a Q of 20.

From the table, we select for $Q=20$ the values

$R_1 = 1\,\Omega,\qquad R_2 = 0.591\,\Omega,\qquad R_3 = 2.693\,\Omega,$

and

$C_1 = C_2 = 1.0000\,\text{F}$

The value of K fixes the gain of the amplifier:

$K = 1 + R_a/R_b$

We choose $R_a = R_b = 1.0000$, giving $K=2$. We now have component values for a filter with the desired Q of 20, but with a centre frequency of 1 radian/second. To obtain the required centre frequency of 5 kHz we divide the capacitor values by $2\pi \times 5000$ and obtain $C_1 = C_2 = 31.8\,\mu\text{F}$. Finally, to obtain reasonable values we multiply all the resistor values by 10 000 and divide all the capacitor values by the same factor. The result is:

$R_a = R_b = R_1 = 10\,\text{K}\Omega,\qquad R_2 = 5.91\,\text{k}\Omega,\qquad R_3 = 26.93\,\text{k}\Omega$

and

$C_1 = C_2 = 3.18\,\text{nF}$

It then remains to select the nearest commercially available values and trim as necessary.

Third-Order Filters

Good characteristics can be obtained for a wide range of applications using third-order low-pass and high-pass filters. These are probably the most popular kind of general-purpose filters, since a third-order filter gives the best performance that can be achieved using a single op-amp.

Third-Order Low-Pass Filters

The circuit diagram of a low-pass filter is shown in figure 4.10. The component values required to achieve particular characteristics are listed in

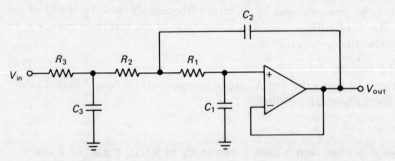

Figure 4.10 Third-order low-pass filter.

ANALOGUE FILTERS 91

Table 4.4

Filter class	R_1	R_2	R_3	C_1	C_2	C_3
Maximally flat (Butterworth) 3.01 dB at ω_2	1.000 00	1.000 00	1.000 00	0.202 45	3.546 5	1.392 6
Equal ripple (Chebyshev) Ripple (dB)						
0.001	1.000 00	1.000 00	1.000 00	0.071 30	2.503 1	0.840 4
0.03	1.000 00	1.000 00	1.000 00	0.077 36	3.312 8	1.032 5
0.10	1.000 00	1.000 00	1.000 00	0.096 91	4.792 1	1.314 5
0.30	1.000 00	1.000 00	1.000 00	0.085 82	7.407 7	1.682 7
1.00	1.000 00	1.000 00	1.000 00	0.058 72	14.784	2.344 4

Table 4.5

Ripple (dB)	Ripple factor (E^2)
0.01	0.002 305 24
0.03	0.006 931 67
0.1	0.023 293
0.3	0.071 519 3
1.0	0.255 925

table 4.4, and the design process involves the use of normalised filters as before. The attenuation for a Butterworth (maximally flat) low-pass filter in the stop band is approximately -18 dB/octave (-60 dB/decade). The exact attenuation of a third-order Butterworth low-pass filter in the stop band is

$$V_2/V_1 = (1+\omega^6)^{1/2} \tag{4.7}$$

The exact attenuation of a third-order Chebyshev (equal ripple) low-pass filter in the stop band is

$$V_2/V_1 = (1+E^2(4\omega^3-3\omega)^2)^{1/2} \tag{4.8}$$

where E^2 is a constant known as the ripple factor which usually lies in the range 0.002–0.3, as shown by table 4.5.

Example

Design a third-order low-pass Chebyshev filter with a ripple of 0.1 dB and a cut-off frequency of 3 kHz.

The filter circuit is as shown in figure 4.10. Component values for the normalised filter are obtained from the table: $R_1 = R_2 = R_3 = 1\,\Omega$; $C_1 = 0.096\,91$ F; $C_2 = 4.7921$ F; $C_3 = 1.3145$ F. To get a cut-off frequency of $2\pi \times 3000 = 1.89 \times 10^4$ radians/second we divide the capacitor values by 1.89×10^4, giving $C_1 = 51.4\,\mu$F, $C_2 = 254\,\mu$F and $C_3 = 70\,\mu$F. Finally, to get realistic component values we divide all the capacitor values by 10^4 and multiply all the resistor values by the same amount, giving $R_1 = R_2 = R_3 = 10\,\text{k}\Omega$, $C_1 = 5.14$ nF, $C_2 = 25.4$ nF and $C_3 = 7$ nF.

The frequency response of the filter can be obtained by using equation (4.8). This equation is, however, for the normalised filter. To use it for a denormalised filter, ω must be replaced with f/f_H where f_H is the passband cut-off frequency. Then

$$\frac{V_2}{V_1} = \left[1 + 0.023\,293\left\{4\left(\frac{f^3}{2.7 \times 10^{10}}\right) - 3\left(\frac{f}{3 \times 10^3}\right)\right\}^2\right]^{-1/2}$$

where $E^2 = 0.023\,293$ from table 4.5 for 0.1 dB ripple. We can now calculate the attenuation at any frequency we choose, and can plot the whole of the frequency response curve for this filter.

Third-Order High-Pass Filters

The circuit for a third-order high-pass filter is shown in figure 4.11. Component values for the design of a third-order normalised high-pass filter are given in table 4.6. For a cut-off frequency of f_L Hz $= 2\pi f_L$ radians/second (that is, when ω_L does not equal 1), the capacitor values must be divided by ω_L as in the low-pass case. To obtain practical component values the resistor values are multiplied by a given factor and the capacitor values divided by the same factor, as in the low-pass case.

The frequency responses for Butterworth and Chebyshev low-pass filters are given by equations (4.7) and (4.8). These equations also give the frequency response for the high-pass case, when the following change is made: ω is replaced by $1/\omega$. For a denormalised high-pass filter, ω is replaced by f_L/f where f_L is the cut-off frequency.

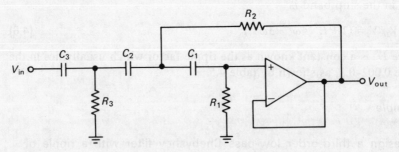

Figure 4.11 Third-order high-pass filter.

Table 4.6

Filter class	R_1	R_2	R_3	C_1	C_2	C_3
Maximally flat (Butterworth) 3.01 dB at ω_1	4.939 49	0.281 94	0.718 08	1.000 00	1.000 00	1.000 00
Equal ripple (Chebyshev) Ripple (dB)						
0.01	10.913 0	0.394 50	1.189 91	1.000 00	1.000 00	1.000 00
0.03	10.097 36	0.301 86	0.968 52	1.000 00	1.000 00	1.000 00
0.10	10.318 8	0.208 68	0.760 75	1.000 00	1.000 00	1.000 00
0.30	11.652 30	0.134 99	0.594 28	1.000 00	1.000 00	1.000 00
1.00	17.029 9	0.067 64	0.426 55	1.000 00	1.000 00	1.000 00

A third-order bandpass filter may be constructed from a filter pair—that is, a low-pass and a high-pass filter in series.

Fourth-Order Filters

Fourth-order filters are designed by cascading (connecting in series) two second-order filters. The circuit diagram of a fourth-order low-pass filter is shown in figure 4.12, and that for a fourth-order high-pass filter in figure 4.13. The component values are determined by using tables 4.7 and 4.8 respectively. These tables list component values for a Butterworth filter, and for a Chebyshev filter with a ripple of 3 dB. Many different combinations of components yield the right characteristics. The tables give two possible

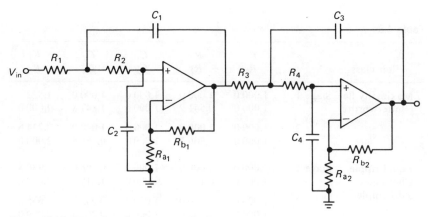

Figure 4.12 Fourth-order low-pass filter.

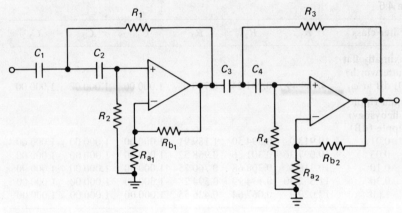

Figure 4.13 Fourth-order high-pass filter.

Table 4.7

Filter class		R_1 R_3	R_2 R_4	C_1 C_3	C_2 C_4	K_1 K_2
Maximally flat (Butterworth)	Stage 1	1.000 0 1.000 0	1.000 0 0.541 2	1.000 0 1.000 0	1.000 0 1.847 8	1.152 2 2.000 0
	2	1.000 0 1.000 0	1.000 0 1.306 5	1.000 0 1.000 0	1.000 0 0.765 4	2.234 6 2.000 0
Equal ripple (Chebyshev) 3 dB ripple	Stage 1	1.000 0 1.000 0	5.102 0 5.122 5	1.000 0 1.000 0	1.000 0 0.496 0	5.004 1 5.000 0
	2	1.000 0 1.000 0	1.107 3 13.603 2	1.000 0 1.000 0	1.000 0 0.081 6	2.198 6 2.000 0

Table 4.8

Filter class		R_1 R_3	R_2 R_4	C_1 C_3	C_2 C_4	K_1 K_2
Maximally flat (Butterworth)	Stage 1	1.000 0 1.000 0	1.000 0 0.541 2	1.000 0 1.000 0	1.000 0 1.847 8	1.522 2.000 0
	2	1.000 0 1.000 0	1.000 0 1.306 5	1.000 0 1.000 0	1.000 0 0.765 4	2.234 6 2.000 0
Equal ripple Chebyshev 3 dB ripple	Stage 1	1.000 0 1.000 0	1.000 0 0.161 3	1.000 0 1.000 0	0.196 0 1.215 2	0.980 8 2.000 0
	2	1.000 0 1.000 0	1.000 0 0.296 0	1.000 0 1.000 0	0.903 1 0.267 3	2.635 8 2.000 0

combinations of components for each stage of each class. Either set of components of the first stage can be combined with either set of components of the second stage of the same class. Components for the first set of each stage are selected so that the normalised values of either both resistors or both capacitors and as many other components as possible are equal to unity. Components for the second set are chosen so that the gain of the amplifier is equal to an integer. This makes it possible to use two resistors of the same value for the amplifier. In particular, a gain of 2 results in normalised resistance values (of R_{a1} and R_{b1} for the first stage and R_{a2} and R_{b2} for the second stage) of unity, since the gain is given by 1 plus the ratio of the resistances. In one case (first stage of the low-pass Chebyshev filter) the gain K_1 is set to 5. A lower value in this case results in negative values of other components and is therefore physically unrealisable.

SPECIAL-PURPOSE FILTER DEVICES

The construction of filters higher than fourth order using discrete op-amps becomes rather complex. In any case it is unnecessary nowadays, since the manufacturers of integrated circuits have thoughtfully provided us (the users) with a variety of special-purpose filter chips of very high performance.

The reader's reaction to this information will probably be something like: 'Well, why bother with all that hard work using op-amps? Let's just use filter chips!'.

However, filters designed around op-amps are, for all their faults, cheap, and are usually good enough for most purposes. A typical filter chip will be at least ten times more expensive than an op-amp filter, and the use of special-purpose devices is really only justified when high performance is needed.

It is not the writer's intention to discuss the inner workings of filter devices, since a very long book could be written on the subject (and many have been). The aim is to describe the specification and use of two typical devices. The interested user will find many more described in the manufacturer's literature.

The first device we shall consider is the RF5609A low-pass filter manufactured by E. G. & G. Reticon, for which a datasheet is given at the end of the chapter. This device is a switched-capacitor filter, with a stop-band roll-off of more than -80 dB/octave as shown on the datasheet. The term 'switched-capacitor' refers to the internal working of the filter, and need not concern us further. The device is completely self-contained, and needs only a power supply and an external clock signal (a regular pulse train) to function. The corner frequency is related to the clock frequency by a fixed ratio (given on the datasheet), which means that the cut-off can be adjusted simply by changing the frequency of the clock pulses. The cut-off can be from 10 Hz to 25 kHz. The ease with which the corner frequency can be changed makes this device specially useful for constructing automatic antialiasing filters for analogue-to-digital conversion systems (see chapter 8), since a multiple of the sample rate can be used to set the cut-off point.

The device will handle signals with a dynamic range of up to 75 dB (dynamic range = amplitude range expressed in dB).

In its internal functioning the device samples the input signal at half the clock frequency. As the ratio of clock frequency to corner frequency is about 100, this means that the input signal is sampled at about 50 times the corner frequency. If the input signal contains frequencies of more than 50 times the corner frequency therefore, it is wise to include an external antialiasing filter on the input. A simple first-order low-pass filter is quite adequate for this, since as the sampling frequency is so much higher than the corner frequency a fairly gentle roll-off is all that is required.

In some applications a residue of the sampling frequency may interfere with the output. If this occurs a simple first-order low-pass filter placed on the output is again usually sufficient to remove the problem. Figure 4.14 shows a complete filter design using the RF5609A.

The second group of filter devices we shall examine are again manufactured by E. G. & G. Reticon, and are bandpass filters. The RM5604, RM5605 and RM5606 contain $\frac{1}{3}$ octave, $\frac{1}{2}$ octave and full octave bandpass filters respectively. The filters are sixth-order Chebyshev class, and like the low-pass device discussed above require only a power supply and a train of clock pulses to function. The filter centre frequency is 1/54th of the clock frequency, and centre frequencies from 10 Hz to 10 kHz can be achieved. (Obviously, the bandwidth is fixed at $\frac{1}{3}$ octave, $\frac{1}{2}$ octave or one octave.) The

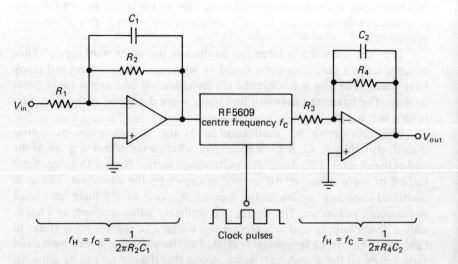

Figure 4.14 Low-pass filter using RF5609, with antialiasing.

ANALOGUE FILTERS   97

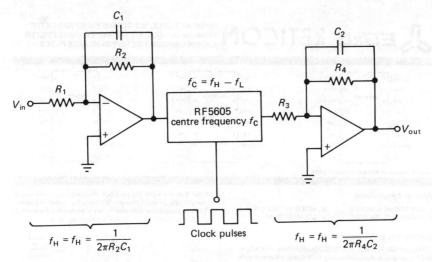

Figure 4.15   Bandpass filter using RM5605, with antialiasing.

device will handle signals with a dynamic range of more than 75 dB and peak-to-peak voltages of 10 V.

Once again this device operates by sampling the input signal at half of the clock rate. If the input signal contains frequency components greater than 27 times the centre frequency therefore, it is a wise precaution to include a simple first-order antialiasing filter on the input as discussed above. Similarly, a low-pass filter on the output may be necessary in some applications to remove residues of the clock frequency. A datasheet for the RM5604, RM5605 and RM5606 is given at the end of this chapter, and figure 4.15 shows an example bandpass filter design using the RM5605.

EG&G RETICON

RM5604A TRIPLE, 1/3 OCTAVE BANDPASS FILTER
RM5605A DOUBLE, 1/2 OCTAVE BANDPASS FILTER
RM5606A SINGLE 1 OCTAVE BANDPASS FILTER

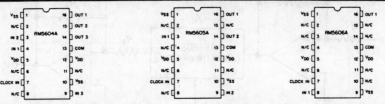

Figure 1. Pinouts for RM5604, RM5605, RM5606

Typical Applications

- SPECTRUM ANALYSIS
- HARMONIC ANALYSIS
- MODEM FILTERING
- EQUALIZATION

Description and Operation

The RM5604, RM5605, and RM5606 contain three 1/3 octave, two 1/2 octave, and one full octave ANSI Class III filters (6-pole Chebyshev) respectively. They are monolithic, switched-capacitor band-pass filters in 16-pin DIP's, requiring only an external clock trigger with plus and minus power supplies to operate. With appropriate pin number changes, the test configuration used to obtain characteristics and operating parameters for the RF5604 applies to the RF5605 and R5606 as well (figure 2).

The center frequency of all filters is tunable by the clock frequency. Dynamic range is better than 75 dB and distortion is less than 0.1%. The filters will handle input signals of greater than 10 volts p-p, and have a typical insertion loss of 0 dB.

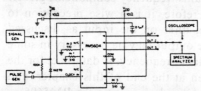

Figure 2. Test circuit for the switched capacitor filters; pin connections shown are those for the RM5604. A similar test circuit applies to the RM5605 and RM5606 filters, with suitable connection changes.

ANTIALIASING CONSIDERATIONS

The sampling rate on the RM5604, RM5605 and RM5606 filters is approximately 54 times the filter center frequency. As in all sampled data systems, signals above half the sampling frequency (fs) will be aliased and may appear in the band of interest. If signals greater than 27 fo will be applied to the filter, an external antialiasing filter may be required. A typical 2-pole antialiasing filter with design equations is shown in figure 3. In applications where the clock residue may affect system performance, a single pole filter should be added to the filter output. Typical clock residue is 25 mV p-p.

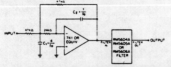

Figure 3. Antialiasing filter for use with RM5604/RM5605/RM5606 filters.

Note that the input trigger-clock frequency, f_c, is twice the sample rate, or approximately 108 times the center frequency. The input clock rate is divided by two in generating the on-chip clock waveforms which control the sample rate, fs.

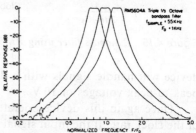

Figure 4: Frequency response RM5604

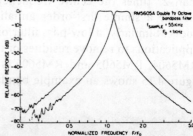

Figure 5: Frequency response RM5605

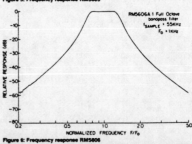

Figure 6: Frequency response RM5606

345 POTRERO AVENUE   SUNNYVALE, CALIFORNIA 94086-4197   (408) 738-4266   TWX: 910-339-9343

ANALOGUE FILTERS

TABLE I: ABSOLUTE MINIMUM/MAXIMUM RATINGS

	Min	Max	Units
Input voltage — any terminal with respect to substrate	-0.4	21	V
Output Short-circuit duration — any terminal	Indefinite		
Operating Temperature	0	70	°C
Storage Temperature	-55	125	°C
Lead Temperature (soldering, 10 sec.)		300	°C

CAUTION: Observe MOS Handling and Operating Procedures

NOTE: Table I shows stress ratings exclusively; functional operation of this product under any conditions beyond those listed under standard operating conditions is not suggested by the table. Permanent damage may result if the device is subject to stresses beyond these absolute min/max values.
Although devices are internally gate-protected to minimize the possibility of static damage, MOS handling precautions should be observed. Do not apply instantaneous supply voltages to the device or insert or remove device from socket while under power. Use decoupling networks to suppress power supply turn-off/on switching transients and ripple. Applying AC signals or clock to device with power off may exceed negative limit.

ANSI S1.11 SPECIFICATIONS FOR TYPICAL 3dB BANDWIDTH AND MINIMUM 40dB ROLLOFF

	FO	3dbBandwidths		40dB Bandwidths	
		LOW	HIGH	LOW	HIGH
Full Octave	1	0.750	1.33	0.240	4.18
Half Octave, Filter 1	1	0.857	1.17	0.438	2.28
Half Octave, Filter 2	1.4	1.20	1.63	0.631	3.19
Third Octave, Filter 1	.8	0.720	.889	0.442	1.45
Third Octave, Filter 2	1	0.900	1.11	0.552	1.81
Third Octave, Filter 3	1.25	1.13	1.39	0.690	2.26

TABLE II: DEVICE CHARACTERISTICS & OPERATION RANGE LIMITS[1] — RM5604, RM5605, RM5606

PARAMETER	CONDITIONS & COMMENTS	SYM	MIN	TYP	MAX	UNITS
Supply Voltages		V_{DD}	+5		+10	V
		V_{SS}	-5		-10	V
Supply Current	RM5604	I_D		16	25	mA
	RM5605	I_D		12	25	mA
	RM5606	I_D		6	10	mA
Input Clock Threshold		V_{IL}	V_{SS}		V_{SS} +0.8	V
		V_{IH}	V_{SS}+2.0		V_{DD}	V
Clock Pulse Width		Tcp	200		T_C -200	nS
Center Frequency		fo	10		10,000	Hz
Clock to Center Freq. Ratio	f_C/f_o RM5604 1st filter		130	136.3	143.5	
	2nd filter		104	109	115	
	3rd filter		83.9	87.2	92	
	RM5605 1st filter		104	109	115	
	2nd filter		73.9	77.4	80.9	
	RM5606		104	109	115	
Q range	RM5604	Q	4.3	4.7	4.9	
	RM5605	Q	3.14	3.18	3.3	
	RM5606	Q	1.72	1.73	1.76	
Input Impedance		R1	3			MΩ
Input Capacitance		C1			20	pF

TABLE III: PERFORMANCE STANDARDS[1] – RM5604, RM5605, RM5606

PARAMETER	CONDITIONS & COMMENTS	SYM	MIN	TYP	MAX	UNITS
Output Noise		EN		0.75	2	mV rms
Dynamic Range		D.R.	73	76	80	dB
Total Harmonic Distortion	1 KHz	T.H.D.		.00	0.3	%
Crosstalk			-66	-70		dB
Passband Ripple				0.20	0.50	dB
Insertion Loss			-0.5	0	0.5	dB
Output Amplitude (2)		Vo	9.5	10		V p-p
Output Current (2)		Io	4		10	mA

(1) V_{DD} = +10V, V_{SS} = -10V, fo = 2 KHz, 25°C (2) Performance degrades at temperatures above 25°C

EG&G RETICON

RF5609A ELLIPTIC LOW-PASS FILTER
RF5613A LINEAR LOW-PASS FILTER

Description

The Reticon RF5609, and RF5613 are monolithic, switched-capacitor low-pass filters fabricated in a double-poly NMOS process.

The RF5609 is a seven-pole, six-zero elliptic low-pass filter with over 75 dB out-of-band rejection and less than ±0.5 dB of passband ripple. The Reticon RF5613 is a linear-phase low-pass filter with over 60 dB out-of-band rejection.

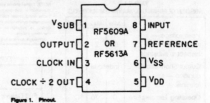

Figure 1. Pinout.

Key Features

- Easy to use
- No external components required
- Small size: 8 pin mini-DIP
- Wide power supply range: ±5V to ±10V
- Dynamic Range: up to 75 dB
- Corner Frequency Range: 10 Hz to 25 KHz
- Insertion loss typically: 0 dB

Typical Applications

- Antialias filters
- Reconstruction filters
- Tracking filters
- Audio analysis
- Telecommunications
- Portable instrumentation
- Biomedical/Geophysical Instrumentation
- Speech processing

Device Operation

The RF5609 and RF5613 are self-contained and require only an external clock trigger, either TTL or CMOS, and power supply. The device characteristic and operating parameters were obtained using the test configuration shown in Figure 2.

In applications where DC information must be passed through the filter, the output offset may be nulled out by varying the reference voltage, which will change the input trigger level and may require adjustment of clock voltage values. The reference input requires less than 100 μA of current and must always be well-filtered. A circuit that may be used to adjust out the output offset is shown as optional resistors in Figure 2.

A divide-by-two clock output is also available. This output contains a square wave at the sample rate and may be used for triggering, summing out the sample rate residue, or driving additional filters especially when filtering requirements are spaced by an octave. Gain and phase tracking from device to device and over the temperature range is typically better than 0.5%. This measurement excludes the fixed offset and fc/fo tolerance at room temperature.

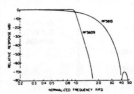

Figure 3. Magnitude response

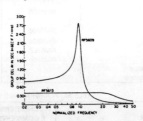

Figure 4. Group delay

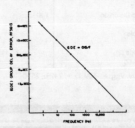

Figure 5. Group delay error
(Second order effects of switched capacitor filter)

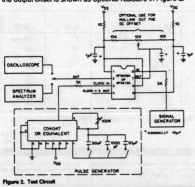

Figure 2. Test Circuit

345 POTRERO AVENUE   SUNNYVALE, CALIFORNIA 94086-4197   (408) 738-4266   TWX: 910-339-9343

ANALOGUE FILTERS 101

TABLE I ABSOLUTE MINIMUM/MAXIMUM RATINGS

	Min	Max	Units
Input voltage — any terminal with respect to substrate	-0.4	21	V
Output Short-circuit duration — any terminal	Indefinite		
Operating Temperature	0	70	°C
Storage Temperature	-55	125	°C
Lead Temperature (soldering, 10 sec.)		300	°C

CAUTION: Observe MOS Handling and Operating Procedures

NOTE: This table shows stress ratings *exclusively*: functional operation of this product under any conditions beyond those listed under standard operating conditions is not suggested by the table. Permanent damage may result if the device is subject to stresses beyond these absolute min/max values. Moreover, reliability may be diminished if the device is run for protracted periods at absolute maximum values.

Although devices are internally gate-protected to minimize the possibility of static damage, MOS handling precautions should be observed. Do not apply instantaneous supply voltages to the device or insert or remove device from socket while under power. Use decoupling networks to suppress power supply turn-off/on switching transients and ripple. Applying AC signals or clock to device with power off may exceed negative limit.

Pre/Post Filtering Considerations

The typical sampling rate on the RF5609 is 50 times the corner frequency; for the RF5613 it is 64 times the corner frequency. (Note: Sampling rate = ½ input clock trigger rate.) Because these sample rates will be far from the frequencies of interest in most cases, antialiasing filtering will usually not be required. However, as with all sampling systems, frequencies or noise above half the sample rate will be aliased and may appear in the band of interest. If this is the case, an external antialiasing filter will be required on the input. A one or two pole Butterworth low-pass filter will usually suffice. An unstable clock frequency can also produce the effect of an aliased signal. In applications where sampling residue may affect system performance, a single pole RC filter may be added to the output.

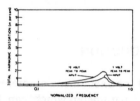

Figure 6. Typical total harmonic distortion

TABLE II: DEVICE CHARACTERISTICS AND OPERATING RANGE LIMITS[1] — RF5609, RF5613

PARAMETER	CONDITIONS & COMMENTS	SYM	MIN	TYP	MAX	UNITS
Supply Voltages		V_{DD}	+5		+10	V
		V_{SS}	-5		-10	V
Quiescent Current (3)	No load	I_Q		12	16	mA
Clock Frequency		fc	1		2500	KHz
Clock Pulse Width		Tc	200	6	10^9/fc-200	nsec
Input Clock Threshold Level		Vth	0.8	2.2	3.0	V
Output signal (2)	Vin = 14-20V P-P	Vo	12	13		V p-p
	RL ≥10KΩ	Io		4		mA
Clock to Corner Freq. Ratio	RF5609	fc/fo	97	100	103	
	RF5613	fc/fo	122	128	134	
Corner Frequency (2)	RF5609	fo	0.01		25	KHz
	RF5613	fo	0.01		19.5	KHz
Input Impedance(s)		Ri		≥10		MΩ
		Ci			≤15	pF
Load Impedance(s)		R_L	≥10			KΩ
		C_L			≤50	pF
Dynamic Output Impedance	Small Signal	Ro		10	250	Ω

TABLE III: PERFORMANCE STANDARDS[1] — RF5609, RF5613

PARAMETER	CONDITIONS COMMENTS	SYM	MIN	TYP	MAX	UNITS
Output Noise		en			2.5	mV rms
Dynamic Range[2]	R5609	D.R.	70	75		dB
	R5613	D.R.	60	65		dB
Total Harmonic Distortion		THD			3.0	%
Insertion Loss[2]				-0.4	0.4	dB
Clock Feedthrough				30	60	mV rms
Ripple				-0.2	0.2	dB
DC Offset Voltage[2]	R5609		-0.6	0.1	0.6	VDC
	R5613		-1.0		1.0	VDC

(1) V_{DD} = +10V, V_{SS} = -10V, fc = 500 KHz, T = 25°C
(2) Performance degrades at temperatures above 25°C
(3) Increases 15% for operation at 0°C; increases 30% for operation at ±10V

055-0032
48261

Chapter 5
Signal Conversion

INTRODUCTION

Most of the phenomena of interest to engineers or scientists occur as changes in some parameter which can be measured over a period of time. Such events are usually measured by sensors which give an analogue voltage or current output. There are of course exceptions. For example, many optical displacement sensors operate by sensing the motion of patterns of dark and light stripes as discussed in chapter 2. The output from such devices may be treated as digital information, although some analogue signal conditioning is often necessary before the data is usable.

Although many sensors provide analogue information describing the state of the measurand, the digital form is without doubt the most convenient in which to analyse and manipulate data. Thus, it is often necessary to convert information from analogue to digital form, or vice versa. The devices used to carry out this task are known as analogue-to-digital converters (ADCs) and digital-to-analogue converters (DACs).

To represent a continuous analogue signal in digital form it is necessary to convert the data at regular intervals. This process is known as sampling or digitisation. A sample rate (in Hertz) is usually quoted rather than the interval between conversions, and care should be taken to select a sample rate which will not cause aliasing problems (as discussed further in chapter 8).

All ADCs and DACs take a finite time to carry out a conversion. If the input signal (in analogue or digital form) changes its value in the midst of a conversion, unpredictable errors may arise. To avoid this effect it is necessary to ensure that for digital-to-analogue conversion the input digital value does not change while a conversion is in progress. For analogue-to-digital conversion special circuits are used, which 'freeze' the analogue input signal during a conversion process. The construction of these 'sample-and-hold' devices will be discussed later in this chapter.

DIGITAL AND ANALOGUE CONVERSION FUNDAMENTALS

Analogue-to-digital (ADC) and digital-to-analogue (DAC) converters have many similarities, and as a class these hybrid analogue/digital devices are frequently referred to as A/D/A converters. An ADC might, for example, be used in digitising analogue data for later use by a computer. A DAC could be used when numbers stored in digital form are to be used to drive the pen on a chart recorder.

The purpose of an A/D/A converter is to associate an analogue voltage, such as 3.7 V, with a binary number such as 00010110. The commonest coding scheme is natural binary, with the MSB (most significant bit) listed first. Figure 5.1 shows the input–output association for a 3-bit ADC with a full-scale voltage of 10 V. (Obviously a 3-bit device is not going to find many applications, but it is simpler to demonstrate the principles involved with only a few bits.)

For a 3-bit device, a total of $2^3 = 8$ output codes are possible. The resolution of an ADC is

Resolution = (Full-scale input)/2^n (5.1)

where n is the number of bits. In the example of figure 5.1 the resolution

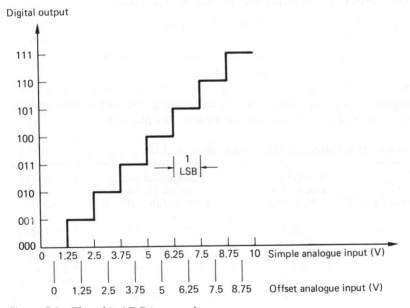

Figure 5.1 Three-bit ADC input and output.

(which is equivalent to the analogue input required to just change the least significant bit or LSB) is 1.25 V. Notice that no digital code corresponds to the full-scale voltage input. This is conventional practice, and has the advantage that converters with identical full-scale voltages but with differing numbers of bits will differ only in resolution; the 'cardinal voltages' such as $\frac{3}{4}$, $\frac{1}{2}$ and $\frac{1}{4}$ of the full-scale output will remain unchanged. Table 5.1 shows that this is so for the $\frac{1}{2}$ full-scale case.

Transitions occur from one digital number to the next at integral multiples of the LSB voltage. By offsetting the input voltage by $+\frac{1}{2}$LSB, the error band may be placed centrally about the integral multiplier of the LSB. For example, if the digital output from the device of figure 5.1 is 100 and the analogue input has no offset, then we can say that the input voltage is 5.00 with a possible error of +1LSB, since the digital output will be 100 for the input range 5.00 to 6.24 V. However, if the input voltage is offset by $+\frac{1}{2}$LSB, then 100 corresponds to $5.00 \pm \frac{1}{2}$LSB. Of course, the total uncertainty is always 1LSB.

Sources of Error

The resolution error of 1LSB ($=FS/2^n$) is an essential quantisation error, and can only be reduced by increasing the number of bits. In practical A/D/A converters additional errors can accumulate due to temperature effects, differential ageing of components etc. An indication of the expected magnitude of these errors is usually given on the datasheet for a particular A/D/A device. These errors may be classified as follows:

Linearity—This is usually specified by the manufacturer of the A/D/A device as the maximum deviation from a straight line drawn between the full-scale output and zero. A good converter should have a maximum non-linearity of no more than $\pm\frac{1}{2}$LSB.

Gain error—Most devices have a nulling input to deal with this effect. However, it is often found that the null setting required is temperature-dependent, and there is no easy solution to this problem.

Table 5.1 10 V Full-Scale ADC, Input Offset by $\frac{1}{2}$LSB

No. of bits	Input change which causes LSB change	Input which will give all 1s out (that is, full scale)	$\frac{1}{2}$ scale
3	1.25 V	8.75 V ± 0.625 V	5 V ± 0.625 V
4	625 mV	9.375 V ± 313 mV	5 V + 313 mV
8	39 mV	9.96 V ± 20 mV	5 V ± 20 mV
10	9.8 mV	9.9902 V ± 5 mV	5 V ± 5 mV
12	2.44 mV	9.9976 V ± 1.2 mV	5 V ± 1.2 mV
16	0.15 mV	9.9999 V ± 0.08 mV	5 V ± 0.08 mV

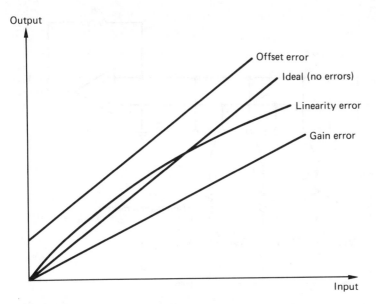

Figure 5.2 A/D/A errors (shown exaggerated).

Offset error—Similar to the above. Once again temperature-dependence can be a problem.

The various errors discussed above are summarised graphically on figure 5.2.

DIGITAL-TO-ANALOGUE CONVERTERS

The DAC has many stand-alone uses, such as the digitally-controlled plotter driver mentioned earlier. However, its primary use is in conjunction with other circuit elements to provide high-accuracy analogue-to-digital conversions. To a large extent the performance of such ADC systems is determined by the performance of the DAC itself.

In chapter 3 we saw a simple form of DAC (figure 3.27) which can be used when only a few bits are required. However, as discussed earlier this kind of DAC becomes unreliable for more than about 6 bits. An alternative arrangement known as the R-$2R$ ladder overcomes this limitation.

In figure 5.3 an R-$2R$ ladder DAC is shown. Switches S_1-S_4 connect the $2R$ resistors (except the termination resistor) either to ground or to the virtual ground of the summing amplifier. The DC currents in the ladder are therefore unaffected by the switch position. To understand the ladder operation, consider a negative reference voltage $-V_{ref}$ attached to the end of the ladder as shown. This will set up DC voltages V_3, V_2 and V_1, and

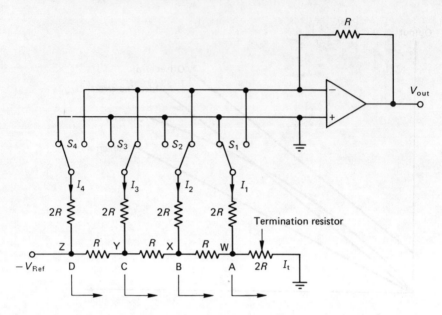

Figure 5.3 R-2R ladder DAC.

DC currents I_4, I_3, I_2 and I_1 will flow. Looking to the right of point A we see that voltage V_1 at node W is applied across a pair of equal $2R$ resistors, and sets up equal currents I_1 and I_t.

The resistance to the right of point B is $2R$ (R in series with R), so with V_2 at node X applied across equal resistors ($2R$ in parallel with $2R$) we can write

$$I_2 = I_1 + I_t$$

but

$$I_1 = I_t$$

so

$$I_2 = 2I_1$$

Now consider point C. The total resistance to the right of C is again $2R$, so I_3 must equal the total current flowing into node Y from the right. Thus

$$I_3 = I_2 + 2I_1$$

but

$$I_2 = 2I_1$$

so

$$I_3 = 4I_1$$

Similarly, the resistance to the right of D is also $2R$, and again I_4 equals the current entering node Z from the right. Therefore

$$I_4 = I_3 + 4I_1$$

but

$$I_3 = 4I_1$$

so

$$I_4 = 8I_1$$

Clearly, a binary weighting scheme has emerged and the ladder may be extended to the left for as many bits as are required. The termination resistor is necessary since it 'fools' the ladder into thinking that the ladder extends infinitely to the right. Because only two resistor values are needed, this scheme can be implemented with the assurance of good temperature tracking.

Digital-to-Analogue Converters—Possible Problems

When the digital input code changes a period of time t_s is required before the analogue output establishes itself at the correct value. The time t_s is called the settling time, and is defined as the time required for the analogue output to settle within $\pm\frac{1}{2}$LSB following the input code change. The settling time requirements are most stringent for the MSB. For example, when switching on the MSB in a 10 V full-scale 8-bit converter the output must settle to within 20 mV of 5 V to achieve 8-bit accuracy. The LSB however needs only to settle to within 20 mV of 40 mV to achieve the same accuracy.

Any circuit used to sense or amplify the digital-to-analogue output can only degrade the settling time, since all circuits contain some capacitance. Special care must be taken to eliminate or reduce as far as possible stray capacitances in circuits connected to the output of a DAC.

For converters with a current output there is a limited range of voltages over which the output can operate at the specified accuracy. This voltage range is called the output voltage compliance, and is expressed in volts above and below the circuit earth voltage.

Spikes (often called 'glitches') in the analogue output can occur at any transition, but they are especially annoying at the cardinal transitions. For example, consider the transition to $\frac{1}{2}$ scale for the 8-bit DAC shown in figure 5.4(a). With the input code at 01111111 all the lower current switches are

108 INSTRUMENTATION FOR ENGINEERS

turned on. When the code advances to the next binary number (that is, 10000000) the lower-order switches may not turn off as fast as the MSB switch turns on. Thus the code 11111111 momentarily appears, causing the inverting buffer amplifier to produce a spike as shown in figure 5.4(a). A sample-and-hold circuit can be used to avoid this effect as shown in figure 5.4(b). The last stable value is held during the digital input transition, and the new value is acquired at least one settling-time later. In the circuit of figure 5.4(b), performance is optimised by making the clock high time t_h slightly greater than the combined settling-times of the digital-to-analogue and buffer amplifier. However, it should be remembered that deglitching is

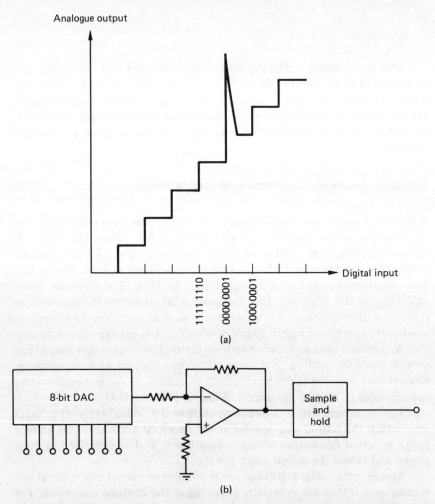

Figure 5.4 DAC output spikes and their remedy: (a) 'glitch' at cardinal transition of an 8-bit DAC; (b) sample-and-hold device used to 'deglitch' a DAC's output.

only achieved at a price, by adding a fixed delay t_h to the overall transfer function of the converter.

FREQUENCY-TO-VOLTAGE CONVERTERS

In the digital-to-analogue conversion applications discussed earlier, the input was assumed to be in binary or some other digital code. However, in some cases the digital input may be a train of pulses or other waveform where the frequency of the incoming signal is used to convey information. In such cases direct conversion to a voltage may be more convenient than counting the pulses, encoding the count as binary and proceeding to use one of the conversion processes discussed in the last section.

In direct frequency-to-voltage conversion, signal conditioning circuits are used to generate a standard pulse for each input cycle. An *RC* low-pass filter and integrator average the pulse train, giving an output voltage proportional to the input frequency.

SAMPLE-AND-HOLD DEVICES

In the introduction we mentioned the need to keep the input to an A/D/A device constant while the process of conversion takes place. With a digital-to-analogue converter this is simply a matter of ensuring that the computer or whatever provides the digital input maintains the binary value for as long as the conversion takes. With ADCs however, the input is normally derived from an external voltage or current source such as a sensor, and it can vary unpredictably. To get around this problem sample-and-hold (S&H) circuits are used, which either pass an input through when set to sample, or 'freeze' it at the ADC's input when on hold.

Many ADCs have built-in S&H circuits. They may also be obtained in integrated circuit form, or can be constructed as follows.

MOSFET switches are the basic ingredients of sample-and-hold circuits. (*Reminder*: A MOSFET or field effect transistor may be made to act rather like a mechanical switch. An analogue signal applied to the input (source or drain) appears at the output (drain or source) un-attenuated as long as the gate terminal is maintained at a positive voltage. When the gate goes below a threshold level, the resistance between input and output becomes very high—typically 10 000 MΩ.) An FET can be used in a sample-and-hold circuit as shown in figure 5.5. The first op-amp (A1) is there to provide a low impedance replica of the input. Notice that it is configured as a unity-gain follower. The FET passes the signal through during 'sample', and disconnects it during 'hold'. Whatever signal was present when the FET is turned off is held on the capacitor *C*. The second op-amp A2 has

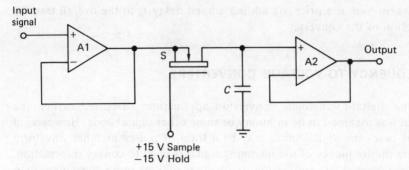

Figure 5.5 Sample-and-hold circuit.

a high input impedance, so the leakage current through the capacitor during hold is very small. The size of the capacitor has to be a compromise: leakage currents flowing back through the FET and through A2 cause the voltage across C to 'droop' during the hold interval, according to

$$dV/dt = I_{\text{leakage}}/C \tag{5.2}$$

Thus, C should be as large as possible to minimise droop. However, the FET's ON resistance (typically 10–50 Ω) forms an RC low-pass filter in combination with C, so C should be small if high-frequency signals are to be followed accurately. Other limitations which must be considered are that amplifier A1 must be able to supply the capacitor's changing current $I = C\,dV/dt$, and that it must have a fast enough slew rate to follow the input signal.

ANALOGUE-TO-DIGITAL CONVERTERS

There are several basic techniques for analogue-to-digital conversion, and the following is a review of the more common approaches. Since users invariably buy a commercial analogue-to-digital chip rather than construct their own circuit, the following descriptions are intended to serve as a guide to aid intelligent selection rather than instructions for building ADCs.

The Parallel Encoder

The parallel encoder (also known as the flash encoder) is conceptually one of the easiest forms of ADC to understand. In a parallel encoder the analogue voltage input is fed simultaneously to one input of each of N operational amplifiers arranged as comparators. The other inputs of the op-amps are connected to N equally spaced reference voltages generated by a resistor string, as shown in figure 5.6. The resistor string sets the inverting input of each comparator at a slightly higher voltage than the inverting input of the

SIGNAL CONVERSION 111

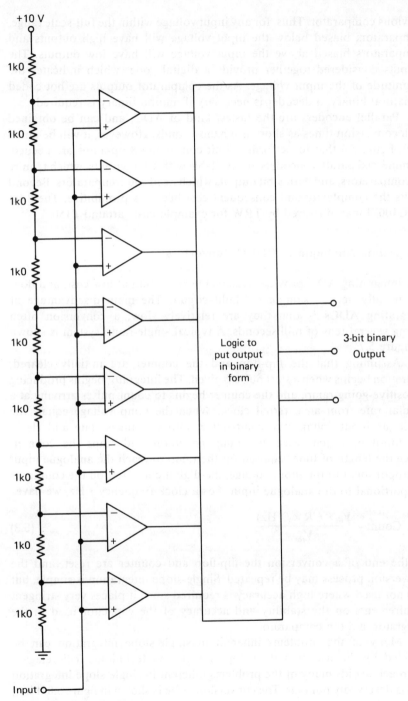

Figure 5.6 Parallel encoder.

previous comparator. Thus, for any input voltage within the full-scale range, comparators biased below the input voltage will have high outputs and comparators biased above the input voltage will have low outputs. The outputs considered together provide a digital code, which indicates the magnitude of the input voltage. As the comparator outputs are not coded in natural binary, a decoder is necessary if natural binary is required.

Parallel encoders are the fastest kind of ADC, and can be obtained with conversion times as short as 6 nanoseconds. However, it will be noted from figure 5.6 that to perform a 3-bit conversion 8 op-amps are needed. Commercial parallel encoders are available with 4-bit outputs, which require 16 comparators, and with 8-bit outputs which need 256 comparators. Beyond 8 bits the complexity and consequent cost becomes prohibitive. The 8-bit TDC1007J manufactured by TRW for example costs around £150.

Integrating Analogue-to-Digital Converters

The integrating ADC provides reasonable resolution at low cost, and does not usually need a sample-and-hold circuit. The main disadvantage of integrating ADCs is that they are relatively slow, a conversion often taking several tens of milliseconds. A typical single-ramp system is shown in figure 5.7.

Assuming that the flip-flop and the counter are initially cleared, operation begins when a start pulse is given. The integrator begins producing a positive-going ramp, and the counter begins to count pulses arriving at a regular rate from an external clock. When the ramp voltage equals the analogue input voltage the comparator output changes state and clears the flip-flop, which resets the ramp generator and stops the counter. Since the length of time required for the ramp to reach the analogue input is proportional to the input voltage, the digital count left on the counter is proportional to the analogue input. For a clock frequency f Hz, we have

$$\text{Count} = \frac{V_{in} \times CR \times f \,(\text{Hz})}{V_{ref}} \tag{5.3}$$

At the end of a conversion the flip-flop and counter are reset, and the conversion process may be repeated. Single-slope integration is simple, but it is not used where high accuracy is required since it places very stringent requirements on the stability and accuracy of the capacitor used in the integrator and the comparator.

Many of the problems inherent in single-slope integration can be avoided by the use of a dual-slope integration technique. This elegant approach avoids many of the problems inherent in single-slope integration and is deservedly popular. The conversion cycle is shown in figure 5.8. First, a current proportional to the unknown input voltage is used to charge a

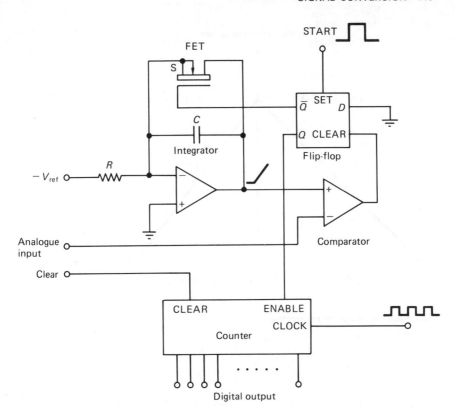

Figure 5.7   Single-slope integrating ADC.

capacitor (which forms part of an op-amp integrator) for a fixed time interval. The capacitor is then discharged at a constant current, until the voltage across it reaches zero again. The time taken to discharge the capacitor is proportional to the input voltage, and can be measured by a clock/counter combination as before. Good accuracy is achieved since the same capacitor is used for both the charge and discharge cycles. Drifts in the comparator are also cancelled out, since each conversion cycle begins and ends at the same voltage.

Successive Approximation (SA) Converters

The successive approximation ADC offers a significant increase in speed compared with the integrating type, for a small increase in cost. Successive approximation converters are probably the most popular type for moderate to high speed applications. In this technique various trial output codes are fed into a digital-to-analogue converter, and the result obtained each time is compared with the analogue input by a comparator as shown by figure

114 INSTRUMENTATION FOR ENGINEERS

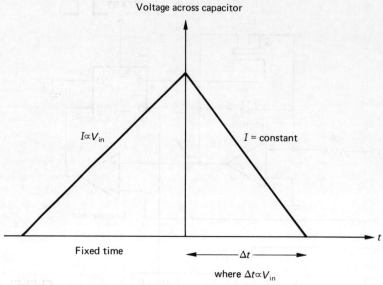

Figure 5.8 Dual-slope integration.

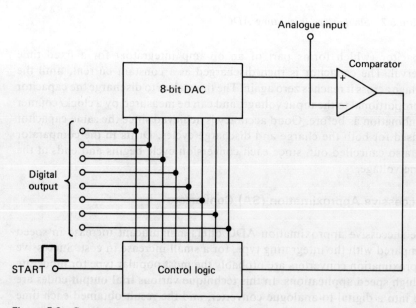

Figure 5.9 Successive approximation ADC.

5.9. The sequence of events is as follows. First, all the DAC input bits are set to 0. Then, beginning with the MSB, each bit in turn is provisionally set to 1. If the output from the DAC does not exceed the input voltage, then that bit is left as a 1; otherwise it is reset to 0. For an n-bit ADC, n such steps are required. An SA converter generally has a 'begin conversion' input and a 'data ready' output. This type of device is relatively accurate and fast, requiring n settling times for the DAC for n-bit precision. Typical conversion times are of the order of a few tens of microseconds.

A variation known as a 'tracking converter' uses an up/down counter to generate the successive trial codes; these are bad at responding to rapid jumps in the input signal but follow smooth changes more rapidly than an SA device. It is essential to use a sample-and-hold circuit on the input of an SA converter.

ANALOGUE MULTIPLEXERS

A digitising system consisting of a sample-and-hold device, an ADC and a computer can only deal with one analogue signal at a time, unless some arrangement is provided which allows a series of signals to be connected to the ADC one at a time. The switching could be done by a mechanical system, but this approach would be impractical for high-speed measurements.

The device universally used to control a number of input signals is known as an analogue multiplexer (often shortened to MPX). These devices are based on FET switches, and use a number of digital inputs (usually coded in binary) to select which of a number of analogue inputs is to be

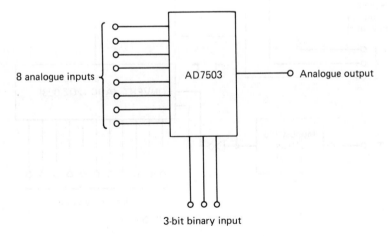

Figure 5.10 Analogue multiplexer.

connected to an analogue output. Figure 5.10 shows a typical design, based upon the Analog Devices AD7503 8-input multiplexer.

EXAMPLE DESIGN

To conclude this chapter we shall examine a typical digitisation circuit design, in which the outputs from a number of sensors are converted to 12-bit digital form. Figure 5.11 shows the circuit, which will digitise the output of any one of eight sensors on command. The sensors produce small signals which have to be amplified or subjected to other signal conditioning as described in chapters 3 and 4.

A 3-bit binary input is used to select one of the multiplexer's eight analogue inputs and connect it to the single analogue output. The MPX output drives a sample-and-hold device, the National Semiconductors LF398. The LF398 output drives the ADC.

The sequence of events during operation of this circuit is as follows. The host device controlling the circuit puts up a channel address to the multiplexer, and gives a START pulse. This triggers a monostable, which generates a SAMPLE pulse. The monostable is used to ensure that the

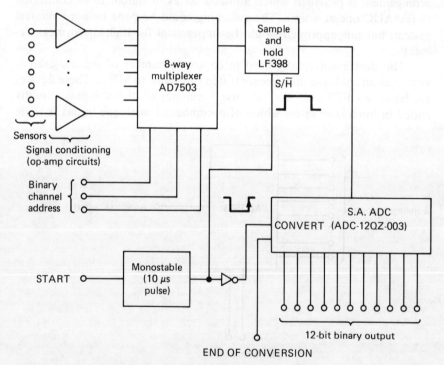

Figure 5.11 Twelve-bit 8-channel digitisation system.

SAMPLE pulse is applied to the sample-and-hold circuit long enough for the S&H's output to settle—typically 10 µs is enough. The monostable's output is also inverted by a NOT gate, and the rising edge is used to give a START CONVERSION signal to the ADC. The host device then monitors the ADC's END OF CONVERSION output, which goes high when the conversion process is complete. The digital output data can then be read from the ADC's output.

Chapter 6
Digital Circuits and Microprocessor Interfacing

INTRODUCTION

In the preceding chapters we have seen how to use sensors, how to amplify or otherwise condition the analogue signal from a sensor, and how to convert data from analogue to digital form or vice versa. Although it is possible to process analogue signals entirely by analogue methods, processing and analysis is much easier if the data is in digital form and is available to a computer or microprocessor. The aim of this chapter is to introduce the subject of digital electronics, and in particular to consider how microprocessor-based computer systems may be interfaced to digital data acquisition and control systems.

An understanding of the structure of microprocessors and how they are programmed is essential when designing interface circuits. Since the subject of this book is instrumentation rather than digital circuit design or microprocessor programming, this chapter only contains an introduction to these topics. The reader who is unfamiliar with these areas is strongly advised to study some of the references in the bibliography for further details, before embarking on any ambitious designs!

Many modern laboratory instruments are provided with a digital interface so that data can be interchanged between a number of instruments, or between an instrument and a computer. There are several 'standard' kinds of digital interface, the most usual being RS232, RS432, IEEE and Centronics. A discussion of these interfaces and how to use them concludes this chapter.

DIGITAL DEVICE FAMILIES

As we have seen, numbers can be represented digitally in binary code using strings of 0s and 1s. This representation is convenient for electric circuits since 0 and 1 can be used to represent the position of a switch, or the presence or absence of a voltage or current.

In addition to facilitating numerical operations the digital form offers the possibility of transmitting data over considerable distances without degradation. This can be done since it is usually possible to distinguish between the presence or absence of a voltage even when large amounts of electrical noise are present. The term 'noise immunity' is used to describe the maximum noise voltages that can be added to logic levels while still maintaining error-free operation.

We usually talk about voltages rather than currents when discussing logic circuits, and call the voltage levels HIGH or LOW. The exact voltages corresponding to HIGH and LOW depend on which particular 'family' of logic devices is being used. The HIGH and LOW states represent the TRUE and FALSE conditions of Boolean algebra. (Boolean algebra is a special kind of mathematics used to analyse combinational logic.) TRUE and FALSE are often written as 0 and 1, which is both a shorthand and a convenient way of indicating that logic levels may be interpreted as binary bits.

Two of the most common 'families' of logic devices are known as TTL (Transistor-Transistor Logic) and CMOS (Complementary symmetry Metal Oxide Semiconductor), after the fabrication process used in each case. Essentially the difference is that TTL devices contain bipolar transistors, whereas CMOS circuits use MOSFETs. In applications where high switching speeds are required, TTL devices should be used. In TTL any voltage between -0.5 V and $+0.4$ V is interpreted as a logic LOW. Any voltage between $+2.4$ V and $+5.5$ V is interpreted as a HIGH. These ranges are deliberately designed into TTL circuits to allow for manufacturing spread, temperature effects on circuits etc. The 74 series of TTL circuits expect their power supply to be within 5 per cent of $+5$ V, and exceeding this range may damage the chip. The behaviour of TTL devices is summarised by table 6.1, which also explains the labelling conventions used.

CMOS devices are designed to tolerate a much wider range of power supply voltages. This is one of the reasons why they are generally preferable to TTL circuits for battery-powered applications. The other reason is that CMOS devices use much less power than TTL, although a penalty is paid in that switching speeds may be much slower. The HIGH and LOW voltages and switching speeds of CMOS devices depend upon the supply voltage, as summarised in table 6.2.

TTL and CMOS devices have different input and output characteristics, which can lead to difficulties when interfacing the two types of device. A TTL input in the LOW state 'sources' (that is, supplies) about 1 mA of current into whatever is driving it LOW. Thus, whatever drives a TTL input LOW must be capable of accepting (sinking) 1 mA. To drive a TTL input

Table 6.1 TTL Device Characteristics and Labelling

TTL DEVICE TYPES

Type	Switching time (ns/gate)	Power dissipation (mW/gate)	Device code
Schottky	3	20	74/54Sxx
Low-power Schottky	5	2	74/54LSxx
Standard*	10	10	74/54xx
Low power*	33	1	74/54Lxx
High speed*	6	23	74/54Hxx

* Rarely used nowadays.

THE 74 AND 54 TTL SERIES

For both the 74 and 54 series of TTL devices the following nominal specifications apply:

Supply voltage	5 V
Logic 0	0.2 V
Logic 1	3 V (minimum)
Noise immunity	1 V

The 54 series is designed to meet a military specification. It will tolerate supply voltages within the range 4.5–5.5 V, and operates over the temperature range −55 to +125°C.

The 74 series is designed to a commercial specification. Supply voltages can be in the range 4.75–5.25 V. The device operates within the temperature range 0–70°C.

DEVICE CODES

TTL devices are labelled as in the following example:

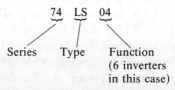

74   LS   04

Series   Type   Function
(6 inverters in this case)

HIGH requires no current in either direction—or to use the jargon, the driving device does not need to source or sink any current. So long as a circuit contains only TTL devices these characteristics present no problems, since the TTL output circuit is good at sinking current. However, CMOS circuits have quite different behaviour, as the inputs require no current at all whether HIGH or LOW and have no source or sink capability. A CMOS output will if required source or sink a limited current, but this is usually less than can be obtained from a TTL device. For a further discussion of the intricacies of TTL/CMOS interfacing see, for example, Horowitz and Hill (full reference given in the bibliography at the end of the book).

Table 6.2 CMOS Device Characteristics

CMOS DEVICES

Recommended power supply range 3–15 V. Switching time, power dissipation and logic levels all depend on supply voltage and level of integration (SSI, MSI, LSI). Typical values in the following table are for SSI devices:

Supply voltage (V)	Switching time (ns/gate)	Power dissipation (nW/gate)	Noise immunity (V)	'0' (V)	'1' (V)
3	50	5	1	<1	>2
5	35	5	1.5	<1.5	>3.5
10	25	10	3	<3	>7
15	22	15	4	<4	>11

If MSI/LSI devices are used, both switching times and power dissipation increase by about a factor of 10.

DEVICE CODES

Most (but not all) CMOS devices are numbered as part of the 4000 series. They are assigned numbers to describe their function—for example, a device labelled 4082 contains two 4-input positive-logic AND gates.

HANDLING PRECAUTIONS

CMOS device inputs are prone to damage from static electricity. You should therefore NEVER touch the input pins with your fingers—the static voltages generated by woollen or nylon clothing can easily destroy CMOS circuitry.

COMBINATIONAL LOGIC, GATES AND BOOLEAN ALGEBRA

The three fundamental logical operations are AND, OR and NOT. All other logical functions can be generated by combinations of these. The output of an AND gate is HIGH (=1) only if all the inputs are high. The behaviour of a gate is usually summarised in a 'truth table'. A two-input AND truth table together with the symbol for an AND gate is shown in figure 6.1. It is important to note that although most books discuss logic gates in their two-input form, gates with larger numbers of inputs may be obtained. In Boolean algebra the AND operation is denoted by a dot. Thus, X AND Y

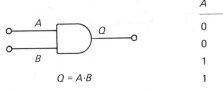

A	B	Q
0	0	0
0	1	0
1	0	0
1	1	1

$Q = A \cdot B$

Figure 6.1 AND gate.

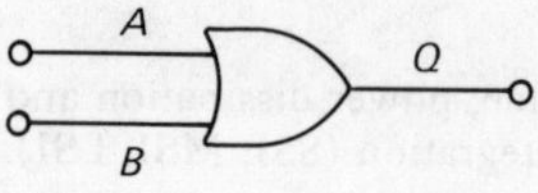

122 INSTRUMENTATION FOR ENGINEERS

A	B	Q
0	0	0
0	1	1
1	0	1
1	1	1

$Q = A + B$

Figure 6.2 OR gate.

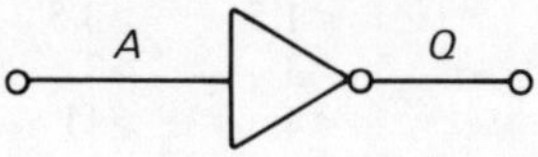

A	Q
0	1
1	0

$Q = \bar{A}$

Figure 6.3 NOT gate.

is written $X \cdot Y$. In some books the dot is omitted so that X AND Y becomes XY.

The output of an OR gate is HIGH if any one or more of the inputs are high. Figure 6.2 shows the symbol for an OR gate together with its truth table. A plus sign is used to denote the OR operation. X or Y is simply written $X + Y$.

It is common in digital circuits to require the complement or inverse of a logic level. This is the function of an inverter or NOT gate. This device can only have one input and one output as shown in figure 6.3. The Boolean symbol for the complement of a logic level is a bar over the symbol. Thus, the complement of A is written $\bar{A}$.

The NOT function can be combined with AND and OR to give NAND and NOR, as shown in figures 6.4 and 6.5.

Exclusive-OR (often written XOR or EOR) is a useful function, although it is less easy to generate from simple gates than NAND or NOR. The output from an exclusive-OR gate is HIGH if one of the inputs is HIGH, and LOW if both inputs are simultaneously HIGH or LOW (it never has more than two inputs). The truth table and circuit symbol are shown in figure 6.6.

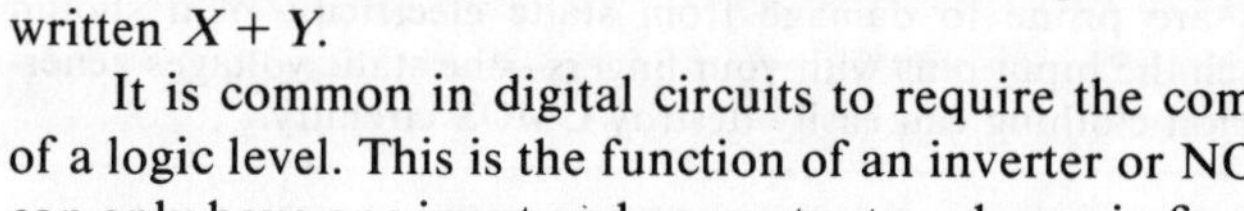

A	B	Q
0	0	1
1	0	1
0	1	1
1	1	0

$Q = \overline{A \cdot B}$

Figure 6.4 NAND gate.

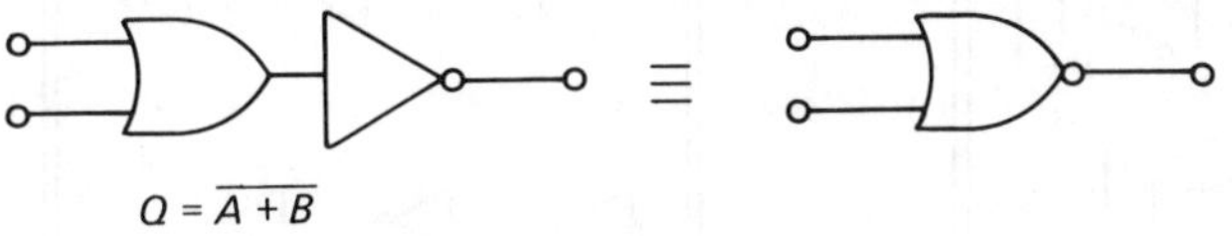

$Q = \overline{A+B}$

A	B	Q
0	0	1
1	0	0
0	1	0
1	1	0

Figure 6.5   NOR gate.

It is seldom necessary to construct logic gates from discrete components, since all the gates you will ever need are already available in integrated circuit form. Figure 6.7 shows a few examples.

To illustrate the use of gates, let us consider a simple example. Suppose we want a circuit which will sound a warning if any door in a car is open, and the driver attempts to start the engine. The circuit is obvious if the problem is restated in Boolean terms as:

Sound buzzer if (Left OR Right door open) AND (Engine started). Expressed symbolically this is

$$Q = (L+R) \cdot E$$

and the circuit (shown in figure 6.8) simply consists of two gates.

The design example above was largely intuitive. With the aid of the logic identities shown in table 6.3 (many of which should be obvious after a moment's thought) logical expressions can be manipulated to find the optimum circuit for a given application. For example, consider the design of an XOR gate. By studying the truth table (figure 6.6) we can see immediately that the output is 1 only when $(A, B) = (1, 0)$ or $(0, 1)$. Thus we can write

$$A \oplus B = \bar{A} \cdot B + A \cdot \bar{B}$$

from which we get the circuit shown in figure 6.9. However, this is not the only way to make an XOR gate. Applying the identities:

$$A \oplus B = A \cdot \bar{A} + A \cdot \bar{B} + B \cdot \bar{A} + B \cdot \bar{B}$$
$$= A \cdot (\bar{A} + \bar{B}) + B \cdot (\bar{A} + \bar{B})$$
$$= A \cdot (\overline{A \cdot B}) + B(\overline{A \cdot B})$$
$$= (A+B) \cdot (\overline{A \cdot B})$$

$A = Q \oplus B$

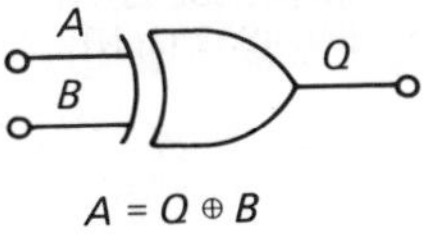

A	B	Q
0	0	0
1	0	1
0	1	1
1	1	0

Figure 6.6   Exclusive-OR gate.

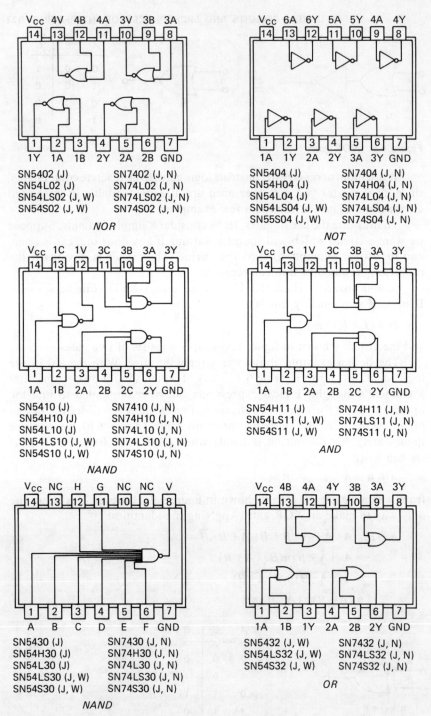

Figure 6.7 Typical example of DIL-packaged logic gates.

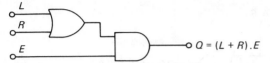

Figure 6.8   Car-door-open warning system.

The final expression leads to the circuit of figure 6.10. By comparing it with figure 6.9 we see that we have saved two logic gates.

SEQUENTIAL LOGIC CIRCUITS

In all the digital circuits we have considered so far, the output(s) at any time have been determined exclusively by the inputs existing at that instant.

Table 6.3   Logic Identities

$A \cdot B \cdot C = (A \cdot B) \cdot C = A \cdot (B \cdot C)$

$A \cdot B = B \cdot A$

$A \cdot A = A$

$A \cdot 1 = A$

$A \cdot 0 = 0$

$A \cdot (B + C) = A \cdot B + A \cdot C$

$A + A \cdot B = A$

$A + B \cdot C = (A + B) \cdot (A + C)$

$A + B + C = (A + B) + C = A + (B + C)$

$A + B = B + A$

$A + A = A$

$A + 1 = 1$

$A + 0 = A$

$\bar{1} = 0$

$\bar{0} = 1$

$A + \bar{A} = 1$

$A \cdot \bar{A} = 0$

$\bar{\bar{A}} = A$

$A + \bar{A} \cdot B = A + B$

$\left. \begin{array}{l} \overline{A + B} = \bar{A} \cdot \bar{B} \\ \overline{A \cdot B} = \bar{A} + \bar{B} \end{array} \right\}$ These two expressions form De Morgan's theorem

---

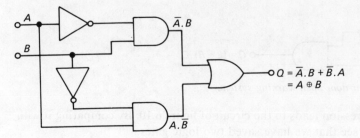

Figure 6.9 Circuit realisation of exclusive-OR.

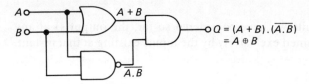

Figure 6.10 Alternative realisation of XOR.

This kind of logic is known as combinational logic, since the output is a Boolean combination of the inputs. Combinational logic circuits have no memory elements, or in other words their present behaviour is unaffected by the history of each input.

Simple gates can be used to construct devices known as flip-flops, which are the fundamental circuit element of sequential logic. The simplest kind of flip-flop (called a reset/set or RS flip-flop) is shown in figure 6.11. To understand its operation, let us begin with $R = S = 1$ and consider the outputs. There are two possibilities; either $Q = 1$ and $P = 0$, or $P = 1$ and $Q = 0$. Thus, we can see immediately that this circuit will adopt either of two stable states. It remains to consider how we can get it to change from one state to the other (that is, flip!).

Suppose we begin with $R = S = 1$, $Q = 1$ and $P = 0$. Now let S be momentarily lowered (that is, S is 1, then 0, then 1 again). When S goes from 1 to 0, P must go from 0 to 1 and therefore Q must go from 1 to 0. The final state is $R = S = 1$, $Q = 0$ and $P = 1$, and we see that the circuit has

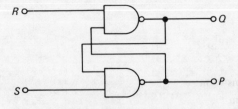

Figure 6.11 RS flip-flop.

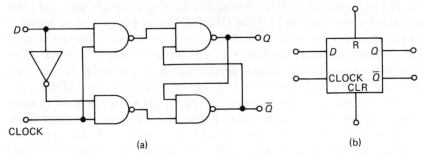

Figure 6.12 The D-type flip-flop: (a) D-type flip-flop using NANDs; (b) D-type flip-flop circuit symbol.

'flipped' into its other stable state. To return to the first state we now have momentarily to lower R.

The RS flip-flop has a number of uses (particularly in switch debouncing). Much more sophisticated flip-flops are available however, ready-packaged as ICs. The two most common ones are the D type and the JK flip-flop.

The D-type flip-flop is made from simple gates as shown in figure 6.12(a). It is usually purchased as a ready-made integrated circuit, and appears as a block on circuit diagrams as shown in figure 6.12(b). The CLOCK input acts as an 'enable' input. When CLOCK is HIGH, any logic level present at D appears immediately at Q. When CLOCK is made LOW the Q output holds the value that D had immediately before the falling CLOCK edge. A $\bar{Q}$ (NOT Q) output is also provided. Many D-type flip-flops also provide RESET (R) and CLEAR (C) inputs, which override the CLOCK and D inputs. Normally R and C are both HIGH, but if R is LOW Q is HIGH and if C is LOW Q is LOW regardless of the other inputs.

The D-type flip-flop can be used to make a register or memory capable of storing a multi-bit binary value as shown in figure 6.13. In a data latch

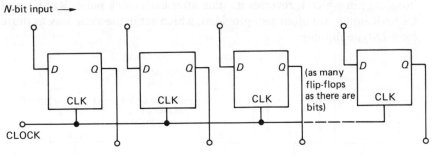

Figure 6.13 D-type register.

(as this arrangement is often known) one flip-flop is used to store the value of each binary digit, and all the CLOCK inputs are connected together. The circuit operates as follows. The value to be stored is presented at all the D inputs in binary, and the CLOCK input is made to undergo either a positive or negative transition (depending on the particular device used). The data on D is then transferred to Q and held until the next CLOCK edge.

A different interconnection of D-type flip-flops may be used to make a shift register. This device has many applications, but one of its most useful functions is to convert data from serial to parallel form or vice versa. The arrangement is shown in figure 6.14, and once again it is usual to obtain these ready-made in integrated circuit form rather than connect them up yourself. The CLOCK inputs are again connected in parallel, but this time each Q output is connected to the next D input. When the CLOCK line is static, the data stored on the Q output of each flip-flop is presented to the D input of the next stage. When the CLOCK undergoes an active transition the data on the D input of each stage is transferred to the Q output, and the effect is to move the whole binary pattern one bit to the right. To convert from serial to parallel form therefore all that is necessary is to present the serial data to the left-hand D input one bit at a time, provide an active CLOCK edge after each bit has settled, and after N CLOCK edges N bits of binary are stored on the N flip-flop. The binary value may then be read in parallel form from the Q outputs of each stage. The reader is invited to deduce how a parallel-to-serial conversion is achieved.

Probably the most widely used kind of flip-flop is the JK type shown in figure 6.15. This figure also shows the 'excitation table' for the device. Many authors call this a truth table, but strictly a truth table may only be used to describe a combinational logic device. The term excitation table is more accurate when dealing with sequential logic. The JK flip-flop is similar to the D type, but there are two data inputs labelled J and K. If we examine the excitation table, we see that if J and K are complements Q will hold the value of J at the next clock edge. If J and K are both LOW, Q is 'frozen'. If J and K are both HIGH, Q will undergo a process known as 'toggling', in which it reverses its state after each clock pulse. RESET and CLEAR inputs are often also provided, which act in the same way as those on a D-type flip-flop.

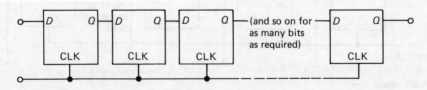

Figure 6.14   Shift register.

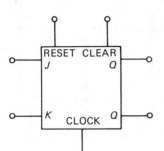

INPUTS					OUTPUTS	
Reset	Clear	Clock	J	K	Q	$\bar{Q}$
0	1	x	x	x	1	0
1	0	x	x	x	0	1
0	0	x	x	x	Undefined*	
1	1	⊓	0	0	Q_{n-1}	$\bar{Q}_{n-1}$
1	1	⊓	1	0	1	0
1	1	⊓	0	1	0	1
1	1	⊓	1	1	Toggle	

Notes: x means 'doesn't matter'.
Q_{n-1} means level of Q before clock edge.
*Reset and clear both low is not an allowed state. The output Q is usually but not reliably high.

Figure 6.15 *JK flip-flop and excitation table.*

The toggling capability of flip-flops can be used to divide the frequency of a train of logic pulses by 2. Figure 6.16 shows the arrangement. Since the flip-flop output changes state on each active clock edge, and there is only one such edge per cycle, the waveform at Q has half the frequency of the CLOCK input.

By cascading several toggling flip-flops (connecting each Q output to the next CLOCK input) it is possible to make a divide by 2^n or binary counter circuit as shown in figure 6.17. However, a major problem with this type of counter is that following input clocking the signal 'ripples' down the chain of flip-flops, momentarily generating incorrect outputs, since it takes each stage a finite time to alter its state. Such a counter is referred to as asynchronous, and can only be used safely for low-speed applications. Synchronous counters are available in which all the flip-flops are clocked simultaneously, which circumvents the problem.

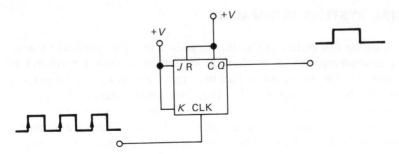

Figure 6.16 *JK used to divide a frequency by 2.*

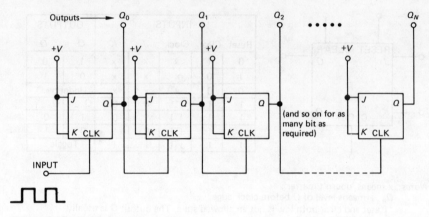

Figure 6.17 Asynchronous counter.

Some counters are provided with a number of control inputs. These make the device much more flexible. Some or all of the following inputs are usually available:

RESET: Almost all counters are provided with a RESET input, which returns the counter output to zero when TRUE.

PRESET: In addition to the output bus an input bus is sometimes provided. The output will assume the value of a binary word placed on this input bus when the PRESET control is TRUE. This is useful if, for example, you do not want to count up from zero but from a pre-defined value, such as 100.

UP/DOWN: Some counters are capable of decrementing as well as incrementing the count each time a clock edge occurs. The direction of counting is controlled by the value of this input.

ENABLE: Some counters have a built-in tri-state buffer on their output bus (see later). When ENABLE is FALSE the output is effectively disconnected, and the count does not appear until ENABLE is TRUE.

DIGITAL SYSTEMS INTERFACING

The following is a review of the devices most commonly used when interfacing microcomputers to mechanical or sensor systems for control or measurement. The list is not exhaustive; there are many special-purpose devices which are not included. It is therefore worth checking that a device tailored to your needs does not exist before embarking on a complex design using simpler components such as gates, latches etc. There are a number of reference books available to assist you in device selection—for example, *The TTL Data Book for Designers* and its companion volume *The CMOS*

DIGITAL CIRCUITS AND MICROPROCESSOR INTERFACING 131

Data Book for Designers, both published by Texas Instruments. These provide details of the functional and mechanical properties of a wide range of devices. It is good practice to examine more than one data book before selecting any components, since data books are usually compiled by device manufacturers and often do not include competitors' products! Flip-flops, counters, shift registers and latches are not included in what follows as these devices have already been discussed.

(a) *Unidirectional buffers.* Also known as *tri-state buffers*, these devices act as switches which logically connect together or disconnect their inputs and outputs. An example is shown in figure 6.18. The ENABLE line acts as the 'finger on the switch'. When it is logically TRUE [which may be high (= logic 1) or low depending on the particular device used] the input is connected to the output. When ENABLE is FALSE a very high resistance is placed between the input and output ports. Effectively, they are no longer connected. This kind of buffer is known as a tri-state buffer, since its output can be in one of three states: HIGH, LOW or disconnected.

These devices are normally packaged in groups of 4, 6, 8 or more to a chip. In some cases each buffer has a separate ENABLE control, but more commonly one ENABLE is used to control all of the data lines at once.

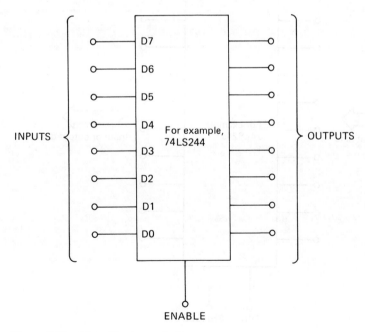

Figure 6.18 *Eight-bit tristate buffer.*

132 INSTRUMENTATION FOR ENGINEERS

Tri-state buffers are used to isolate and protect circuits from each other. If a number of subsystems are connected to a common bus, it is necessary to arrange matters so that only one device at a time places data on to the bus. Similarly, only one device at a time is usually required to respond to data placed on the bus. Tri-state buffers perform this function, the ENABLE line to each buffer being selected by means of a separate address bus.

It is also usual to protect digital lines which are carried between systems using wires (for example, printer cables) by buffers, to prevent circuit damage if the cable is inadvertently cut or disconnected.

Tri-state outputs are available on a wide variety of devices in addition to buffers, and they are invaluable in designing interconnection circuits.

(b) *Bidirectional buffers*, or *bus transceivers*. These act in a similar fashion to the unidirectional buffers described above, but with one important difference: instead of one side always being the input and the other always the output, the flow of data can take place in either direction. As shown by figure 6.19, a second control input is added to define the direction of data flow. At any time the flow of data may be turned on or off by means of ENABLE, and the direction of flow controlled by

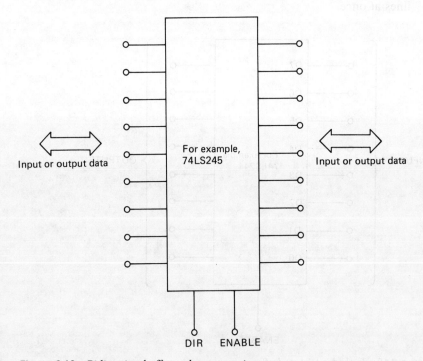

Figure 6.19 *Bidirection buffer or bus transceiver.*

the DIR line. As before, these devices are used for device/bus isolation and for circuit protection, in cases where data input and output is required on the same bus. Also as before, bus transceivers are normally packaged in groups of 4, 6 or 8 to a device.

(c) *Address decoders.* It is common to want to enable or disable buffers by means of addresses specified from a microprocessor program. Address decoders perform the function of decoding a binary address (represented by the state of *n* address lines), and using it to select a single output line for control purposes. The commonest type is the '3-to-8 line decoder', which uses a 3-bit address to choose which one of 8 output lines it makes TRUE. Other sizes are also available. An ENABLE input is also usually supplied, so that the device can be rendered unresponsive when necessary. This ENABLE input can be driven in turn by other address decoders, thus allowing complex addressing structures to be built up.

The output line selected by the binary address assumes a logically TRUE state as long as the address is maintained on the input. This TRUE state may be a zero (0 volts) or a one (+volts), depending on whether the device operates on positive or negative logic. The unselected outputs remain FALSE. A typical 3-to-8 line decoder is shown in figure 6.20.

(d) *Comparators.* As the name implies, these devices compare two sets of binary inputs. An output is produced indicating whether the two binary values are equal, or if they are not the direction of the inequality. As shown on figure 6.21, the input data buses are normally labelled *A* and *B*. Outputs are provided which indicate whether $A = B$, $A < B$ or $A > B$.

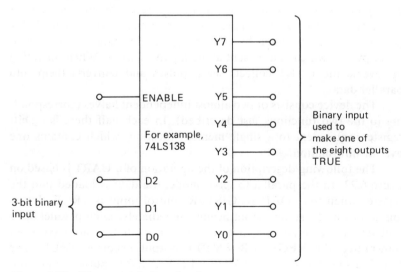

Figure 6.20 Three-to-eight line decoder.

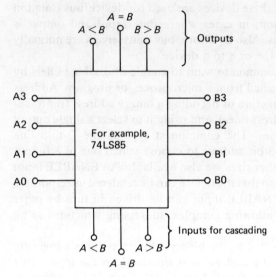

Figure 6.21 Four-bit magnitude comparator.

Inputs (referred to as cascade inputs) are also provided, labelled $A = B$, $A < B$ and $A > B$. These are for connecting (cascading) a number of 4-bit comparators together, so that comparison of larger bytes (more bits) can be achieved. Do not confuse the digital device described here with the analogue comparator discussed in chapter 3.

(e) *The Universal Asynchronous Receiver/Transmitter (UART)*. When it is desired to send data through cables over more than about a metre, it is very inconvenient to have to use a separate conductor for each bit of the bus, which may be as much as 32 or 64 bits in width. The UART gets around this problem by converting parallel data to a stream of serial pulses, which can be sent along a pair of wires. When operating in reverse the UART collects serial pulses and converts them into parallel data.

The device consists of two almost independent halves (corresponding to the two functions just described). In each half there is a shift register, equivalent to a single memory location, which contains one byte of information.

The following description of the operation of a UART is based on figure 6.22. In the parallel-to-serial mode, the data is loaded into the register when the LOAD REGISTER control input is TRUE. Transmission from the serial output begins immediately. Each parallel bit is sent down the serial line in turn, and when all the bits have been transmitted the REGISTER EMPTY output becomes TRUE. The UART also sends DATA START and DATA STOP pulses before and after serial transmission, for use by the receiving device.

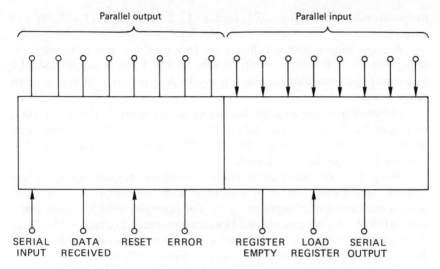

Figure 6.22 Eight-bit Universal Asynchronous Receiver/Transmitter (UART).

In the serial-to-parallel conversion mode, the register is loaded bit by bit as the serial data is received. When the expected number of bits of data have been loaded, plus the correct DATA START and DATA STOP control pulses, the DATA RECEIVED output goes true, telling the host device that valid data is waiting to be read. This is usually achieved by ENABLEing a built-in tri-state output buffer.

The above description outlines the usefulness of the UART. However, a penalty has to be paid in terms of operating speed if one is used within a system. It is obviously much faster to transmit parallel data than serial, and the use of UARTs will almost always slow down the operation of a digital system.

NUMBER CODES

In the discussion so far we have been dealing either with logic levels, which are self-explanatory, or with natural binary numbers. How a digital level can be used to represent numerical values is an interesting and involved question.

In the familiar decimal system a number is represented by the value and position of a digit. The number 23.7, for example, means $(2 \times 10^1) + (3 \times 10^0) + (7 \times 10^{-1})$. In words we say that the number base is 10, and each position to the left or right of the decimal point corresponds to a power of 10.

In binary arithmetic there are only two possibilities, 0 and 1. The base is 2, and just as in the decimal case the position of a digit represents a particular power of 2. Thus, 101_2 means $(1\times 2^2)+(0\times 2^1)+(1\times 2^0)$, or 5 in base 10.

A single binary digit is called a bit. In a digital system information is represented by a sequence of bits. A sequence of 8 bits is usually called a byte (although some authors call it a word). A sequence of 16 bits is often called a word.

In binary notation an n-bit data word can represent 2^n possible values, so long as the values are either all positive or all negative. In microprocessor arithmetic the most significant bit (the MSB) is often used as a sign bit, and is 1 if the number is negative.

The process of signed arithmetic in a computer is based upon the fact that adding the complement of a number is the equivalent of subtraction when a fixed number of digits are used. For example, consider the decimal expression $(5-2)$. We can add the 10's complement of 2, which is $10-2=8$. We then have $5+8=13$, which can be considered as a carry and a 3. By discarding the carry we obtain the correct answer, which is 3.

The system used in microprocessors is similar except that it is carried out in binary. One way to define a negative number $(-N)$ is to say that it is the value which when added to a positive number of the same magnitude $(+N)$ will give zero. In a system with a limited number of digits such as a microprocessor, every number has its negative. For example, working with four bits:

```
        0001  +
        1111
(1)     0000
```

Since we are restricted to four bits the carry is lost from the end, and we see that $-1_{10} = 1111_2$. It can easily be shown that $1110_2 = -2_{10}$, $1101_2 = -3_{10}$, $1100_2 = -4_{10}$, and so on. With this representation adding and subtracting signed numbers is trivially easy, since they are *all* added without taking any account of the sign! For example, the subtraction $5-2$ now becomes:

```
        0101  +
        1110
(1)     0011
```

Once again the carry bit is lost from the end of the sum. In most microprocessors a special bit (known as a CARRY FLAG) is used to indicate that this has happened.

The only drawback of the system outlined above is its potential ambiguity, as there is no way of telling whether, for example, 1100 means $+12$ or -4. The answer to this difficulty is essentially a matter of convention. The usual arrangement is best illustrated by means of a circle diagram.

If we are dealing with 8-bit binary, we can represent 256 ($=2^8$) possible values from 00000000 to 11111111. Adding 00000001 to 11111111 will result

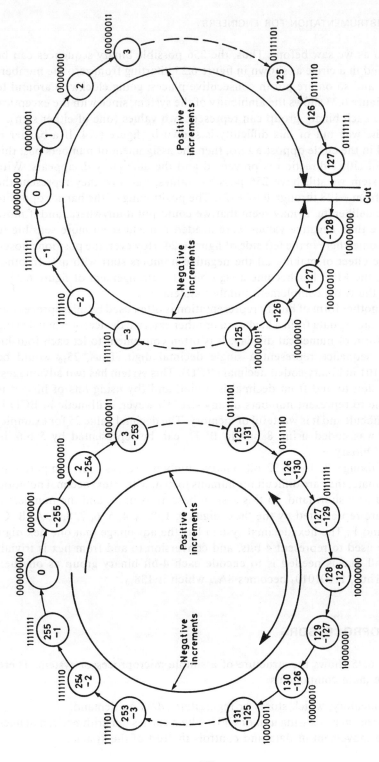

Figure 6.24

Figure 6.23

(*Reproduced from Programming for Microprocessors by A. Colin, by permission of the publishers, Butterworths & Co.*)

in zero as we saw before. Thus, the 256 possible binary sequences can be arranged in a circle as shown in figure 6.23. Starting from zero the numbers 1, 2, 3 and so on are put in consecutive places, going clockwise around to 255. Figure 6.23 shows the ambiguity of the system, since with the exception of zero each binary group can represent two values (one of either sign).

The way out of this difficulty is shown by figure 6.24. If a barrier is placed in the circle opposite zero, then the assignment of numbers past this point in either direction is prevented and the ambiguity disappears. With this system we still have 256 possible values, but now they represent the range from -128 through 0 to $+127$. The positioning of the barrier opposite zero is deliberate. It may seem that we could put it anywhere, and if more positive than negative values were needed it might seem more sensible to put it somewhere in the left side of figure 6.24. However, the position chosen has the effect of making all the negative numbers start with a 1. In other words, the MSB has become a sign bit. The arrangement of figure 6.24 is called the two's complement number system.

Another form of binary representation is often used by microprocessors when reading data from a keyboard or other external device, or when driving some form of numerical display. It is often convenient to let each four-bit binary sequence represent a single decimal digit. Thus, 25_{10} would be 0010 0101 in binary coded decimal (BCD). This system has two advantages: conversion to and from decimal is trivial, and (by using lots of bits) it is possible to represent numbers of any size. However, arithmetic in BCD is very difficult and it is wasteful of memory. The decimal value 25 for example, which was coded using 8 bits in BCD, can be represented by 5 bits in natural binary.

Although 8, 16 or 32 bit binary numbers are easy for computers to manipulate, they are difficult for humans to deal with. Hexadecimal notation is used as a shorthand. In this system base 16 is used, and the numbers 0 to 15 are represented by the 'hex' digits 0, 1, 2, 3, 4, 5, 6, 7, 8, 9, A, B, C, D, E and F. The hexadecimal system has the advantage that one hex digit can be used to represent 4 bits, and conversion to and from hex is trivial, since all that is needed is to encode each 4-bit binary group as one hex digit. Thus, 11001010_2 becomes $8A_{16}$ which is 138_{10}.

MICROPROCESSORS

Figure 6.25 shows the structure of a simple microprocessor system. There are five main components:

(a) A memory, which stores or regurgitates data on demand.
(b) A central processing unit or CPU, which carries out arithmetic, manages the movement of data and controls the rest of the system.

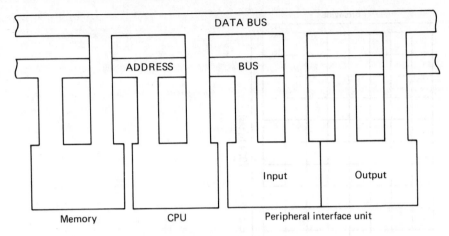

Figure 6.25 Structure of a simple microprocessor.

(c) One or more communication devices which allow access to the outside world.
(d) A data bus which allows data to be sent to all of the other parts of the system.
(e) An address bus, which allows the CPU to alert the individual components of the system to the fact that they are required to take some action.

In small computer systems these five parts may be fabricated on one integrated circuit, in which case the result is called a microprocessor. Larger computer systems may use separate devices for some or all of these functions.

The Memory

The memory of a microprocessor can be thought of as a rectangular array of bits as shown in figure 6.26. Each row of bits forms a byte or word, and is always handled as a single entity. The number of memory locations available varies between a few hundred and several million, but is always a power of 2 (since each location's address is in binary).

Computer memory comes in two fundamental types, known as RAM and ROM. RAM or random access memory can accept data sent to it for storage, and will reproduce it later on demand. ROM or read-only memory cannot accept data, but reproduces permanently stored information on demand. The contents of ROM are usually fixed by the designer of the system.

To read the contents of a memory location the CPU has to put the address of that location on the address bus. If we take the Rockwell 6502 microprocessor as an example, the data bus is 8 bits wide and the address bus is 16 bits wide. Thus, the CPU can address $2^{16} = 65\,536$ memory locations.

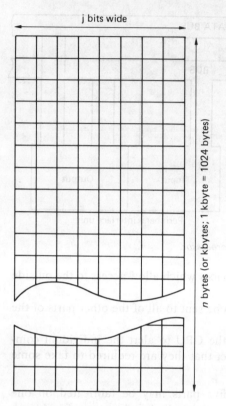

Figure 6.26 Memory organisation.

In the 6502 the instruction LDA $C050 is used to read the contents of location $C050_{16} = 49232_{10}$ into the accumulator, which is the central register used for arithmetic. The $ prefix before the address is conventionally used to denote a hexadecimal number.

To write data into a memory location a similar sequence of events has to happen. The CPU must put the address of the required memory location on the address bus, and then put the data to be stored on the data bus. Using the 6502 as an example again, suppose we want to store the number $A9_{16}$ in location $5000. Two instructions are necessary:

LDA#$A9 (which puts the value A9 into the accumulator). [Note the # prefix in addition to the $, which indicates that A9 is a value, not an address.]

STA $5000 (which stores the accumulator contents—A9—in location $5000).

The CPU

The CPU of a microprocessor has the task of fetching instructions and their associated data from memory, and sequentially executing the instructions. Figure 6.27 shows a hypothetical simple CPU containing four registers and some additional arithmetic and control circuits. The CPU is driven by a series of clock pulses, each of which may produce some change in the register contents.

The memory associated with our simple microprocessor consists of 65 536 (2^{16}) locations, each of which is capable of storing an 8-bit value (a byte). Thus, any memory location can be specified by a 16-bit address, and the program counter register is 16 bits wide.

An operating cycle begins when the CPU is ready to fetch an instruction from the memory and execute it. At this point the program counter contains the address of the next instruction to be fetched and obeyed. The clock pulse causes the program counter contents to be placed on the address bus, which has the effect of selecting one memory location. The contents of that location are sent along the data bus and placed in the current instruction register.

An instruction consists of a binary sequence which has to be decoded by the CPU before it can be executed. The instruction format, which is a basic property of the CPU design, will be something like this:

    ffffffff   aaaaaaaa aaaaaaaa

    FUNCTION   16-BIT ADDRESS

The function specifies the action to be taken by the CPU, and the address usually indicates which memory location is to be used in doing it. The address part of an instruction can in some circumstances contain numerical values.

The function is specified by an 8-bit sequence as shown above. This means that our microprocessor can have 256 possible instructions, each of which is 'seen' by the CPU as an 8-bit binary number. Humans find

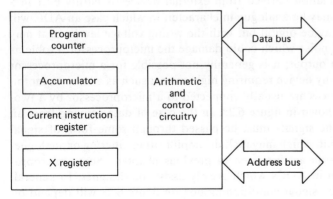

Figure 6.27 Simple microprocessor CPU.

142 INSTRUMENTATION FOR ENGINEERS

instructions in binary difficult to deal with, so each instruction is represented by a 3-letter mnemonic, such as LDA for 'Load into the Accumulator' which we saw earlier.

The third register in the CPU (see figure 6.27) is called the accumulator, and is used in much the same way as the display on a calculator to hold the partial results of calculations. The accumulator is simply a special register in which arithmetic can take place. For example, the pair of 6502 instructions

LDA #$0A
ADC $4000

puts 10 in the accumulator ($10_{10}=0A_{16}$), and then adds the contents of address $4000 to the accumulator. The accumulator ends up containing a value which is the sum of the contents of address $4000 and the value 10.

Most CPUs contain additional registers which can be used for temporarily storing intermediate results, in the same way as the memory on a pocket calculator. The 6502, for example, has two extra registers besides the accumulator, called the X and Y registers. Many of the operations that can be done in the accumulator can also take place in the registers. In our simple hypothetical microprocessor there is one such register, shown as the X register on figure 6.27.

The Peripheral Interface

Inside a microprocessor data is handled as binary sequences of a fixed length. In our simple example each group or byte is 8 bits long. Physically, the voltages representing these 8 bits are arranged to suit the CMOS technology used to make microprocessors. However, at some point it is usually necessary for the microprocessor to communicate with the outside world, and information derived from external devices is hardly ever in a suitable form. It may be analogue in character, in which case an ADC will be required. It may be digital but with the wrong voltage levels, or it may contain electrical noise which could damage the microprocessor. Problems can also occur in output; it is generally not possible for a microprocessor to drive directly any device requiring high current such as a lamp or motor.

External devices are usually connected to a microprocessor by a two-stage system as shown in figure 6.28. In the case of data arriving from an external device the signals must be passed through some form of signal conditioning circuit, which may include amplification, filtering or analogue-to-digital conversion as discussed in previous chapters. Some microprocessors have built-in ADCs which greatly assist the designer. In general, the contents of the signal conditioning box on figure 6.28 will depend on the circumstances.

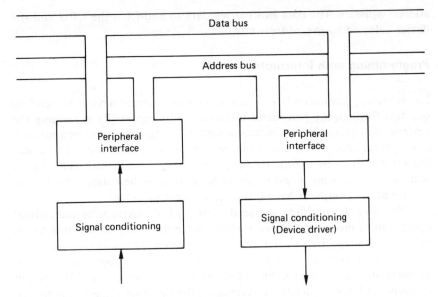

Figure 6.28 Microprocessor peripheral interface unit.

The peripheral interface is used to connect the conditioned signal to the data bus of the microprocessor. It ensures that the data is available when required, and protects the microprocessor system from the inadvertent application of electrical noise, high-voltage spikes etc. A peripheral interface unit has an associated address or range of addresses, and acts very like a memory location in that it will only place data on or accept data from the data bus when its own address appears on the address bus. Thus for many applications the programming associated with peripheral control consists simply of read or write instructions to the peripheral interface, which the CPU treats just like any other memory location.

The output side of figure 6.28 works in a similar way. Output data is placed in a register in the peripheral interface unit, and is converted for use by outside devices by the device driver. The device driver's design depends on the particular application, but typically it may contain a DAC, a stepper motor driver, or transistor switches used as lamp or solenoid drivers.

The Data and Address Buses

From the discussion above it should by now have become apparent that all the parts of a microcomputer, or indeed of any computer, are linked by both an address and a data bus. All the constituent parts of a microprocessor 'listen' to the address bus continuously, and respond only when their own address appears. The data bus carries data to and from the CPU and the device selected by the address bus.

Programming with Interrupts

For many applications it is sufficient to control external devices by sending out data through the peripheral interface, or by repeatedly examining the peripheral interface for any incoming data. With this programming approach a peripheral cannot 'tell' the computer when something needs to be done, but must wait to be asked. However, the problem with this technique is that devices requiring a fast response have to have their status checked so often that the computer has no time to do anything else.

To solve the problem we need a method of interrupting the normal operations of the microprocessor when (and only when) some action needs to be taken. A special connection called the interrupt line is used, which is connected to all the external devices and to the microprocessor. If an external device generates an interrupt (for example, an ADC might announce it has completed a conversion), the interrupt input to the microprocessor is made TRUE. The CPU's circuits spot this and cause normal program execution to be suspended. The microprocessor puts the program counter contents into a special store, so that it knows where to resume its original activity after the interrupt has been dealt with, and jumps to a special (reserved) memory location. The value contained in that location gives the start address of a special program (known as the interrupt service routine or ISR) which takes appropriate action and finishes by resetting the interrupt input. Once the ISR has been completed, normal program execution is resumed.

EXAMPLE INTERFACE DESIGNS

This section contains two example designs in which peripheral devices (a pulse counter and a multiplexed ADC) are connected to the peripheral interface unit of a microcomputer. Both of the examples have been constructed and run from an APPLE II microcomputer; but the design principles involved are quite general and can be applied to other microprocessor-based computer systems.

Example Design No. 1: Two-Channel Pulse Counter

The problem is as follows: you are required to design an interface to allow TTL pulses from a pair of conveyor-belt package detectors to be counted and the total read by a microprocessor. The 6502

DIGITAL CIRCUITS AND MICROPROCESSOR INTERFACING 145

microprocessor has a 16-bit address and an 8-bit data bus. The memory map allows external devices to be connected in the address range $C100–$C1FF.

In the problem definition we are told that our particular microprocessor sets aside the area of memory with addresses in the range $C100–$C1FF for use by external devices. In practice this means that addresses in this range are not ROM or RAM locations, but correspond to external devices connected by means of a peripheral interface.

The first stage in designing any circuit connected to a microprocessor is to arrange for the circuit to recognise its own address from among all those appearing on the address bus. This process is known as address decoding. In general, some combination of comparators and 3-to-8 line decoders can be used to fulfil this function.

A suitable address recognition/decoding circuit is shown in figure 6.29. Remembering that we want the circuit to recognise and respond to addresses of the form $C1xx (where x denotes don't care), the 8 most significant bits of the address bus are connected to one side of an 8-bit comparator (two cascaded 4-bit devices to give an 8-bit comparison). The other side of the comparator circuit is 'hard-wired' to appropriate voltage levels (logic 0s and 1s) to represent in hexadecimal $C and $1. When any address beginning $C1xx appears on the address bus, an $A = B$ signal is generated by the cascaded comparators, which is used to switch on the rest of the circuit.

In this case, the rest of the address recognition circuit simply consists of a 3-to-8 line decoder. The least significant three bits of the address bus are decoded by this device and used to select one of the eight output lines. This gives us a maximum of 8 address-selectable control lines, which is ample for this circuit.

Note that it may be necessary to place an inverter in the connection between the comparator output and the 3-to-8 line decoder. This is a common technique which has to be used when, as in this case, a TRUE:HIGH output from the comparator is to be used to control a TRUE:LOW input such as the 74LS138 ENABLE. Usually it causes no problems since inverters operate very fast (about 15 ns for the 74LS04). However, it should be remembered that device delays are cumulative, and inserting too many devices into one arm of a dual–parallel system can cause a timing imbalance. In such cases one remedy is to insert pairs of inverters into the faster side of the system to provide a delay.

Note also that in our application five bits of the address bus are unused. It is good practice however to connect all the address

146 INSTRUMENTATION FOR ENGINEERS

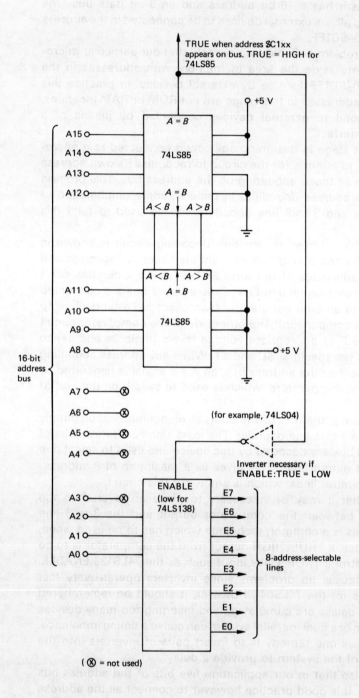

Figure 6.29 Hard-wired address recognition cct

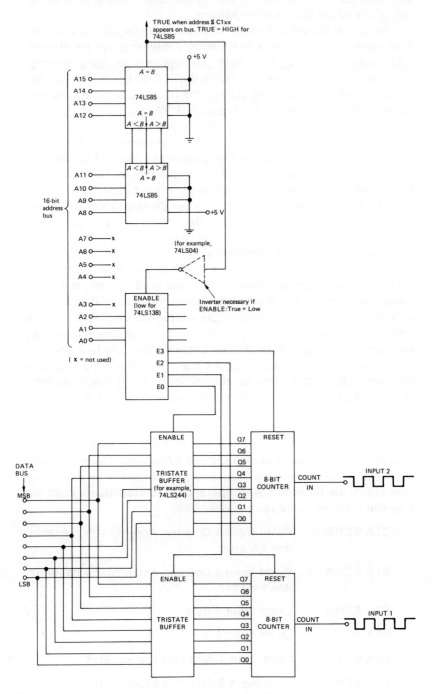

Figure 6.30 Two-channel pulse counter, interfaced to 8-bit microprocessor.

bus lines to terminations on the circuit board whether used or unused, to allow for future expansion.

Having designed our address decoding circuit we can now use it to control the conveyor-belt package-counting system. Figure 6.30 shows how this is done. Each package as it passes a detector produces a TTL pulse, which is fed to the CLOCK input of a counter. Since the total is to be read out and the counters reset at frequent intervals, 8-bit devices (with a capacity of 255) are sufficiently large, and conveniently match the size of the data bus.

The counter outputs cannot be connected directly to the microcomputer data bus, since the changing data values which appear as counting proceeds would inhibit the operation of the computer. Buffers have to be used therefore as shown in figure 6.30.

The two counter RESETS and buffer ENABLES are driven by the 3-to-8 line decoder outputs. Depending on the particular counters and buffers used, it may be necessary to place inverters in some or all of these control lines.

The circuit as drawn in figure 6.30 has one flaw: unless the conveyor belt is stopped during each data read, the count will be continuously changing. If a counter is incremented while its buffer is ENABLED, the resultant bit changes will probably cause erroneous data to be read by the microprocessor. To avoid this happening, we interpose *D*-type data latches between the counters and the output buffers as shown in figure 6.31. Two further 3-to-8 line decoder outputs are used to control (or CLOCK) these latches, and between clock operations static data is therefore held ready by the latch for onward transmission by the buffer to the microprocessor.

The software necessary to control a circuit of this kind can be written in a high-level language such as BASIC (for instance, using PEEK and POKE statements), but would more usually be written in assembly code for high-speed operation. The statements used by a typical machine-code routine would be:

```
STA $C105      Reset Input 1 Counter. Accumulator contents
               don't matter

STA $C104      Reset Input 2 Counter. Accumulator contents
               don't matter

STA $C103      Latch Input 2 data

STA $C102      Latch Input 1 data

LDA $C101      Load Input 1 data into accumulator

LDA $C100      Load Input 2 data into accumulator
```

DIGITAL CIRCUITS AND MICROPROCESSOR INTERFACING   149

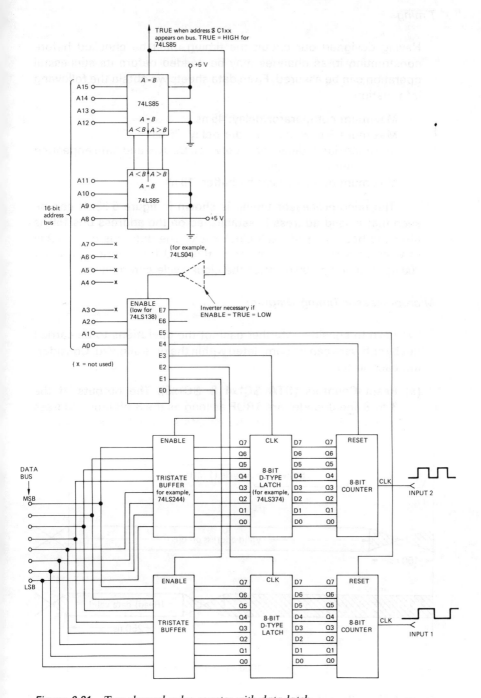

Figure 6.31   Two-channel pulse counter with data latch.

Timing

Having designed our circuit the timing must be checked before constructing it, as changes may be needed before its successful operation can be assured. From data sheets we obtain the following information:

Maximum comparator delay: 45 ns
Maximum 3-to-8 line decoder delay: 25 ns
Maximum latch delay (time between CLOCK and data appearing at Q outputs): 25 ns
Maximum delay caused by buffer: 18 ns

The microprocessor timing is shown in figure 6.32. It can be seen that a valid address is established on the address bus 150 ns after the beginning of each clock cycle. The data bus is available to write data to or read data from the CPU for a period of about 200 ns, beginning 700 ns after the clock cycle commences.

Microprocessor Timing Diagram

We have to consider whether each of the operations to be carried out by software can be completed within the time allowed. Considering each in turn:

(a) Reset Counters (STA $C1x4 & $C1x5). The outputs of the 3-to-8 line decoder are TRUE as long as the 3-bit input address is maintained. However, the counters are edge-reset, so the

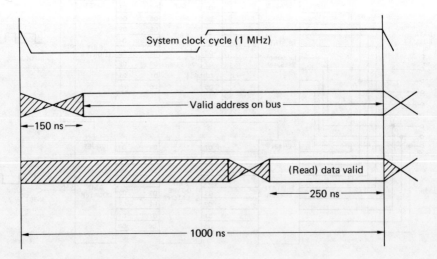

Figure 6.32 *6502 read cycle timing.*

question of timing does not really arise. In general, counter resets are unlikely to be time-critical operations as they are usually part of initialising the system prior to a run.
(b) Clock Latches (STA $C1x3 & $C1x2). Once again, these devices are edge-triggered, so timing problems are unlikely to arise.
(c) Read Data (LDA $C1x0 & $C1x1). This is the area where timing problems are most likely to occur, since we have to ensure that the output data appears on the data bus at the time of the READ DATA window. The data buffers are ENABLED by their addresses (for example, $C100). Each address is maintained on the address bus by a command such as LDA $C100 for 850 ns, starting 150 ns after each clock cycle as shown by figure 6.32. The data buffers will be ENABLED for the whole of that time, and will therefore be holding the data on the data bus once the buffer's output delay (18 ns) has occurred. The data will therefore be maintained on the address bus throughout the DATA READ window (see figure 6.33). Our circuit is satisfactory therefore.

Example Design No. 2: 8-Input Multiplexed ADC for 6502-Based Microcomputer

In this case we are designing an analogue input unit for a computer using a 6502 microprocessor. The analogue input is to be multiplexed so that up to eight sensors at a time can be connected to the

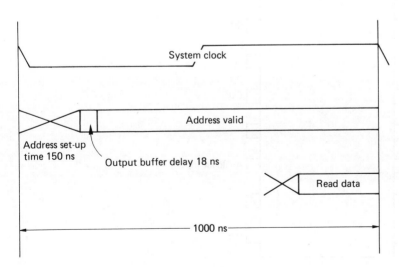

Figure 6.33 Timing diagram for counter circuit.

system. The sample rate is to be initially determined by software, but once set is to be maintained by hardware (so that the microprocessor is free to carry out other tasks). The longest sampling interval required is 10 seconds. The multiplexer is to be advanced from one input to the next automatically.

Datasheet Information

(a) The A-to-D converter shown in figure 6.34 has an 8-bit output and is of the successive-approximation type. The maximum conversion time is 25 µs. In addition to the analogue input there are two digital inputs, and an output which indicates the device's status.

The START CONVERSION (SC) input initiates the conversion process. It is triggered by a positive edge.

The ADC has a built-in tri-state buffer for output control, and the OUTPUT ENABLE (OE) input is the ENABLE for this buffer. When OE is HIGH no data appears at the 8-bit output; when OE is LOW data is placed on the 8-bit bus.

An EOC (End of Conversion) output is provided in addition to the data output. This is provided so that the outside world, or in our case the controlling microprocessor, knows what is going on inside the ADC. EOC is HIGH before a conversion,

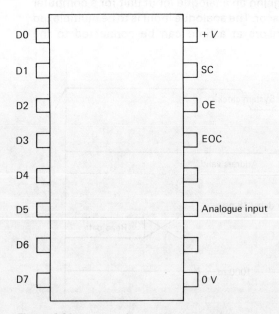

Figure 6.34   Pin-out diagram for 8-bit ADC.

becomes LOW as soon as an SC pulse is received, remains LOW while the conversion is taking place, and returns to HIGH only when the conversion is finished and the binary value is ready for use.

(b) The multiplexer shown in figure 6.35 uses a 3-bit binary address to select one of eight analogue inputs and connect it to a single analogue output. The resistance between a selected input and the output is 30 Ω. The maximum time taken to switch inputs is 200 μs. The multiplexer also has a built-in sample-and-hold circuit, controlled by the S/H input.

Summary of Design Points

(1) The microcomputer is address-mapped for peripheral expansion as in the previous example. The same address decoding circuit is used therefore.

(2) The SC pulses are to be generated at a fixed rate by the circuit (and not by the CPU). We therefore need some sort of timing circuit. Simple timing circuits can be based on counters and comparators as shown in figure 6.36. The principle of operation is that data is written to latches by the CPU, and the latched value is compared with the value reached by a counter counting pulses from a regular source such as the system clock. Each

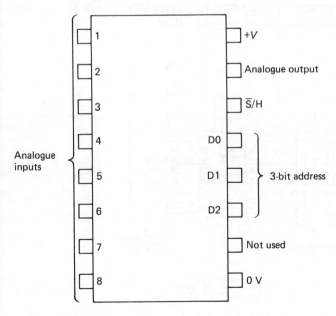

Figure 6.35 Pin-out diagram for 8-input multiplexer.

154 INSTRUMENTATION FOR ENGINEERS

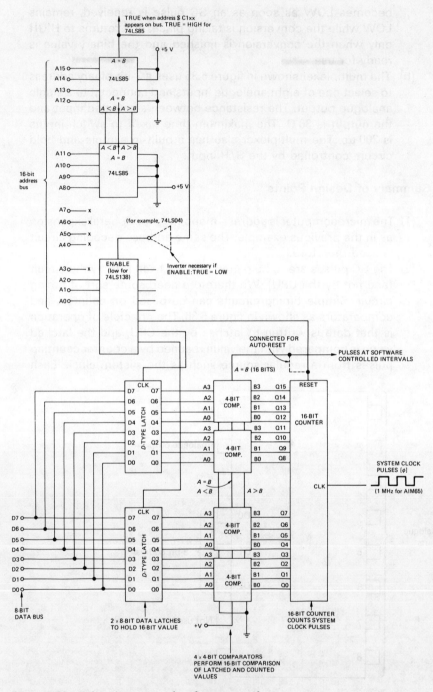

Figure 6.36 Pulse generator with software control.

DIGITAL CIRCUITS AND MICROPROCESSOR INTERFACING 155

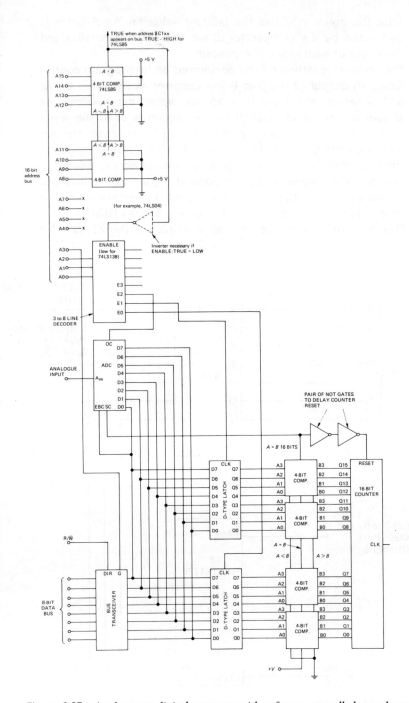

Figure 6.37 *Analogue-to-digital converter with software-controlled sample rate.*

156 INSTRUMENTATION FOR ENGINEERS

time the count matches the latched value an $A = B$ pulse is generated by the comparator chain, the counter is reset and the cycle of operations is repeated.

(3) The counter resetting can be performed under program control using an output of the 3-to-8 line decoder as shown in figure 6.36. Alternatively, it can be made to happen automatically as shown by the dotted line. If this approach is used the $A = B$ pulse resets the counter. However, it may be necessary to use a monostable to ensure that the $A = B$ pulse is long enough for the rest of the circuit, since the instant it appears it resets the counter and consequently disappears! Figure 6.37 shows the circuit with the ADC in place, and a pair of inverters is used to delay the counter reset slightly after $A = B$ goes HIGH.

(4) The automatic advance of the multiplexer (MPX) can be achieved by connecting the MPX 3-bit address port to a 3-bit

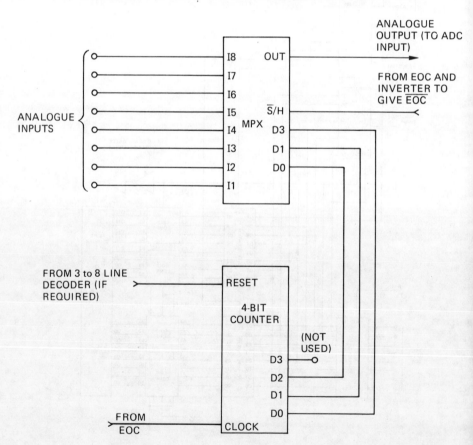

Figure 6.38 Multiplexing circuit.

counter and using the positive-going edge of EOC as the CLOCK input. Each successful conversion will then advance the MPX to its next input.

(5) The MPX sample-and-hold input has to be driven HIGH to hold the data while the ADC performs a conversion. The inverse of EOC, $\overline{EOC}$, can be used for this purpose. Figure 6.38 shows the multiplexing circuit.

COMMUNICATION STANDARDS

Twisted-pair wire conductors or coaxial cables are used to send digital signals to and from a remote device. ASCII (or any other alphanumeric binary code) can be transmitted as a parallel 8-bit group, or as a serial string of 8 bits sent one after another. An ASCII keyboard usually generates an 8-bit parallel TTL-compatible output, but for reasons of convenience serial format ASCII is used to send data to and from remote terminals.

One popular serial data interchange standard is known as the RS-232 standard. The data is sent asynchronously, with a start bit and a stop bit (or sometimes 2 stop bits) attached to the ends of each 8-bit character code, forming a 10 or 11 bit group. The sender and receiver use a fixed bit rate, which may be 110, 150, 300, 600, 1200, 2400, 4800, 9600 or 19 200 baud. (A baud is a bit per second. The rate at which data is transmitted depends on the format used: 8-bit ASCII code is usual, together with a start bit and one or more stop bits.) The RS232C standard defines the voltage levels used as shown in figure 6.39. A second standard for serial data transmission, known as RS-423, is capable of being driven over longer distances. The maximum range of an RS-232 cable link depends to some extent on local factors such as electrical noise, required baud rate etc, but is around a hundred metres. For transmission over longer distances either a lower baud rate or a repeater or both may be necessary.

In addition to the pair of wires used for data transmission the RS-232C standard specifies a number of extra connections for two-way coordination or checking. This process is known as handshaking. The number of handshake connections implemented varies from installation to installation, and can range from none to as many as 17.

For high-speed parallel data transmission over ranges up to about 15 metres the most common arrangement is the IEEE-488 instrumentation bus.

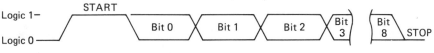

Figure 6.39 RS232 serial data byte timing.

158 INSTRUMENTATION FOR ENGINEERS

This was first defined by the Hewlett-Packard company, and was subsequently used as the basis of an international standard by the Institution of Electrical and Electronic Engineers in the USA. It is sometimes referred to by its earlier name as the Hewlett-Packard Interface Bus or HPIB.

IEEE data links are commonly used for controlling instrumentation. A number of IEEE devices can be connected to an IEEE bus, and signal their status to a host controller which issues commands for the transfer of data. Many manufacturers of laboratory instruments have followed Hewlett-Packard's lead in providing their products with IEEE interfaces, which makes the setting up of complex laboratory experiments fairly simple.

Each device connected to an IEEE bus has an address, which is often switch-selectable, and the controller of the bus issues commands which are prefixed with the address of the device concerned. Obviously it is up to the user to make sure that no two devices on the bus have the same address, or confusion ensues!

IEEE devices are divided into talkers, listeners and controllers. Normally there is only one controller, and the other devices are instructed to behave as talkers or listeners as appropriate. The usual arrangement is to instruct all the devices to act as listeners until they are called, when they become talkers and disgorge their data.

Chapter 7
Frequency Domain Analysis

INTRODUCTION

The analysis of electrical signals is a fundamental problem for most engineers. As we have seen, even if the basic problem is not electrical, the parameters of interest are usually changed from pressure, force, temperature etc. into an electrical quantity by means of transducers.

So far we have made the implicit assumption that we are treating signals as records of the history of some parameter. In other words, we are examining the behaviour of the variable quantity we are interested in as time passes. This approach is known as working in the time domain.

However, this is not always the best approach to use. For many applications there are advantages to be had from examining the behaviour of a parameter with respect to frequency rather than time. This is known as working in the frequency domain. A frequency domain representation of a signal can take several forms, ranging from simple plots of amplitude or phase against frequency to more complex representations such as modal response and waterfall diagrams.

The traditional way of observing signals is to examine them in the time domain. As we saw earlier, the time domain consists of records of the behaviour of a parameter of the system as time passes. For instance, figure 7.1 shows a simple spring-mass system with a pen attached to the mass. A piece of paper is pulled past the pen at a constant rate. The result is a graph of displacement against time, or in other words a time domain view of the displacement of the system.

The French mathematician Fourier showed that any periodic waveform is equivalent to the sum of a number of sinusoids. By selecting the right amplitudes, frequencies and phases any periodic waveform may be synthesised. For example, the waveform of figure 7.2 is produced by the addition

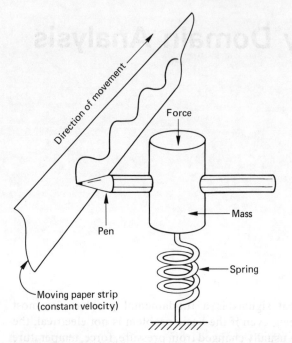

Figure 7.1 *Displacement of spring-mass system as a function of time.*

of a pair of sinusoids. The time and frequency domain representations of this addition are obtained by viewing the problem from two different perspectives, as shown by figure 7.3(a). Two of the axes are time and amplitude, just as we had when working in the time domain. The third axis is frequency, which allows the sinusoidal components of the waveform to be visually separated. If this three-dimensional graph is viewed by looking along the frequency axis, the representation shown in figure 7.3(b) is obtained. This is the time domain view of the waveform, obtained by adding together the component sinusoids at each moment in time.

However, if we look along the time axis as in figure 7.3(c), a different view appears. The axes are now amplitude and frequency, and the signal is observed in the frequency domain. Each sinusoidal component of the complex waveform appears as a vertical line. The height of each line

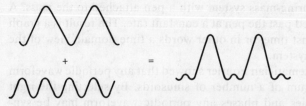

Figure 7.2 *Equivalence of periodic waveform to a sum of sinusoids.*

FREQUENCY DOMAIN ANALYSIS 161

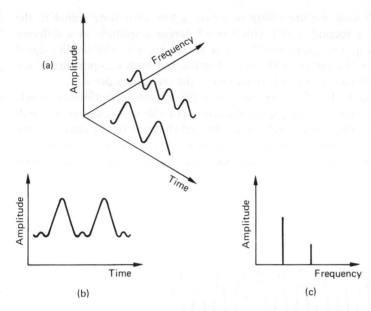

Figure 7.3 (a) Time and frequency domains represented in three dimensions. (b) Time domain view. (c) Frequency domain view.

represents the amplitude of that component, and the line's position indicates its frequency. Since each line represents a sinusoid, this representation characterises the waveform in the frequency domain. This representation is known as the amplitude spectrum of the waveform, and (apart from the phase of each component, which will be discussed later) it contains all the information necessary to construct the complex waveform from its sinusoidal components. It is important to see that we have neither gained nor lost information; we are just looking at it differently. We are looking at the same three-dimensional graph from two different viewpoints.

Since no extra information is provided by examining a signal in the frequency domain, and since the engineer requiring data in this form usually has to go to some lengths to get it, the reader may well wonder whether it is worth the trouble. The following examples will illustrate the utility of the method.

We might want to measure the level of distortion in the signal from an electronic oscillator. The oscillator is designed to produce a sinusoid at a fixed frequency, but the output is distorted by the addition of a number of other frequencies with small amplitudes.

Alternatively, suppose it is required to monitor a bearing and detect the first signs of its failing in a noisy machine. Once again, the problem is to detect a low-amplitude signal in the presence of a large amplitude signal at a different frequency.

162 INSTRUMENTATION FOR ENGINEERS

In each case we are trying to detect a low amplitude signal in the presence of a second signal with a much larger amplitude at a different frequency. Figures 7.4(a) and (b) show what happens—the smaller signal is masked by the larger in the time domain, but both components of the signal appear clearly when it is viewed in the frequency domain.

Let us consider the appearance of a few common continuous signals in both the time and frequency domains. (In the next chapter we shall examine the effect on a spectrum of the fact that practical signals are not continuous. Because data collection apparatus is always turned on at some time and switched off some time later, real signals always have a finite length.)

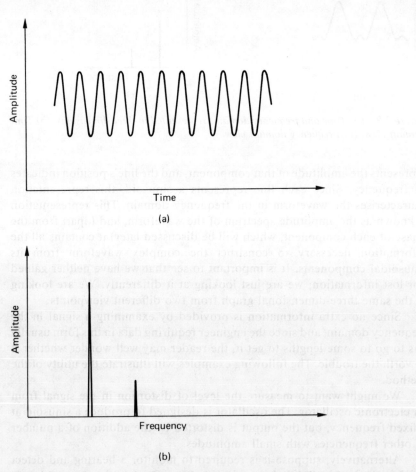

Figure 7.4 The use of the frequency domain to study small-amplitude signals close to larger amplitudes: (a) in the time domain a small signal is masked by a larger one; (b) both signals can be clearly seen in the frequency domain.

FREQUENCY DOMAIN ANALYSIS 163

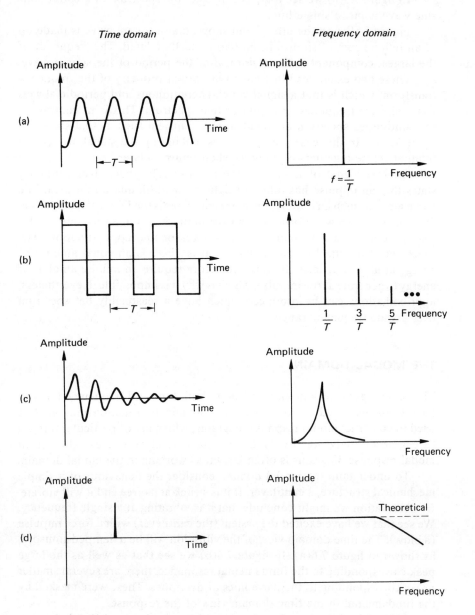

Figure 7.5 Signals in the time and frequency domains: (a) sine wave; (b) square wave; (c) transient; (d) impulse.

In figure 7.5(a) we see that, as expected, the spectrum of a continuous sine wave is just a single line.

Figure 7.5(b) on the other hand shows that a square wave is made up of an infinite sum of sinusoids, all harmonically related. The frequency of the largest component is the reciprocal of the period of the square wave.

These two examples illustrate an important property of the frequency transform, which is that a signal which is continuous and periodic always has a discrete frequency spectrum or line spectrum. This is in contrast to the continuous spectrum obtained from a transient such as that shown in figure 7.5(c). In this case the sine waves making up the signal are spaced infinitely close together—a point we shall return to later.

Another signal of interest is the impulse. The theoretical definition states that an impulse has infinite height, zero width and a unit area. The name given to such an impulse by mathematicians is the Dirac delta function. Obviously, such a waveform cannot exist in practice, but a very sharp pulse such as that shown in figure 7.5(d) is a reasonable approximation. The spectrum of a delta function is flat—that is, there is an equal amount of energy at all frequencies. It would therefore require an infinite amount of energy to generate a true impulse, which is why it is impossible. Nevertheless, approximations can be produced which have a reasonably flat spectrum over a wide frequency range.

THE MODAL DOMAIN

The advantages of examining a signal in both time and frequency domains have been discussed above. Frequency domain information may also be used to construct modal response diagrams, which are of particular interest to engineers analysing the dynamic behaviour of structures. The use of modal response diagrams is often known as working in the modal domain.

To understand the modal domain consider the behaviour of a simple mechanical structure, a cantilever. If it is struck at its free end it will vibrate. By observation we might conclude that it is vibrating at a single frequency. We see that we have excited the system (the cantilever) with a force impulse (a blow). The time domain view of the vibration will be a damped sinusoid, as shown in figure 7.6(a). In figure 7.6(b) we see that as well as the large peak corresponding to the fundamental resonance, there are several smaller peaks corresponding to the harmonics or overtones. These were masked by the fundamental in the time domain view of the response.

Each of the frequency domain peaks corresponds to a mode of vibration of the cantilever. In this case we expect from structural dynamic considerations that the largest peak will be caused by a mode shape like that shown in figure 7.7(a). The second harmonic will be caused by the mode shape shown in figure 7.7(b).

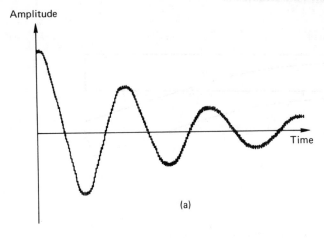

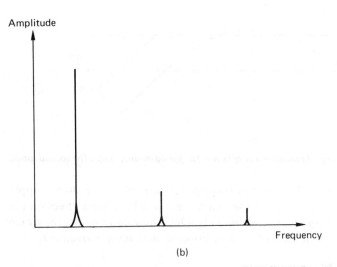

Figure 7.6 Time and frequency domain views of the vibrations of a cantilever: (a) time domain view; (b) frequency domain view.

The vibration of any structure can be represented just as validly by a sum of its modes of vibration as by a sum of its constituent frequencies of vibration. The task of modal analysis is to determine the shape and magnitude of the structural deformation in each vibration mode. Once this has been done, the overall dynamic behaviour of the structure may be modified by changing the individual modes.

In the case of the cantilever beam, we might for example decide that it was unsatisfactory as a structure because the amplitude of the second mode was too large. To reduce the contribution made by this mode, we might measure the vibration and determine that the modes of vibration

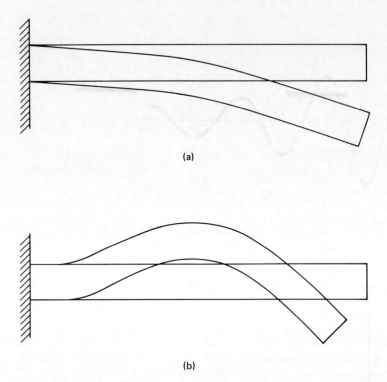

Figure 7.7 Modes of vibration of a cantilever: (a) fundamental mode; (b) second mode.

were those of figure 7.7. The solution then becomes clear: we should apply extra damping to the centre of the beam. This will have little effect on the first (fundamental) mode, but will greatly affect the second mode, for which the damping has been applied at a position of maximum movement.

Modal Domain Measurements

In the example above the dynamic characteristics of a cantilever were modified after its mode shapes had been measured. To determine the dynamic behaviour of such a structure, its amplitude of vibration must be measured at several points. When the resulting sets of time domain data are transformed into the frequency domain, they can be displayed in the form of a modal response diagram such as that shown in figure 7.8. Notice that the third axis indicates the position used for measurement. If the vibration is linear, the sharp peaks (the resonances) all occur at the same frequencies regardless of where they were measured on the structure. By measuring the width of each peak (the Q factor), it can also be shown that the damping of each resonance is independent of the position at which it is measured. The only parameter that varies from position to position is

FREQUENCY DOMAIN ANALYSIS 167

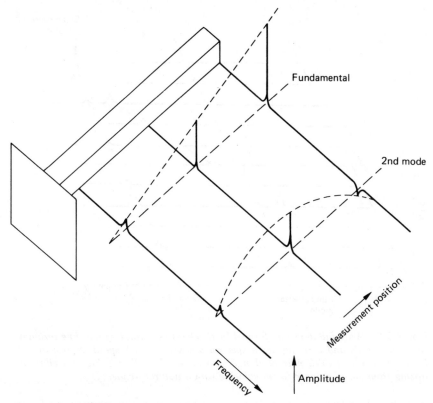

Figure 7.8 Modal response diagram.

the height of the peaks, reflecting the amplitude of each mode at each point. By fitting a line to the peaks of a given mode, the displaced shape of that mode is reproduced. A three-dimensional representation of amplitude, frequency and measurement position, such as that shown in figure 7.8 for the cantilever, gives a clear picture of the displaced shape of a structure for each mode of vibration.

WATERFALL DIAGRAMS

The waterfall diagram or Campbell diagram is used to study the behaviour of rotating machinery. A waterfall diagram may be thought of as a modal response diagram in which the axis representing position is replaced by one representing rotation rate. A rotating machine is accelerated or decelerated slowly through a range of speeds, and spectral analysis is carried out at each speed. The results are plotted as shown in figure 7.9. It can be seen

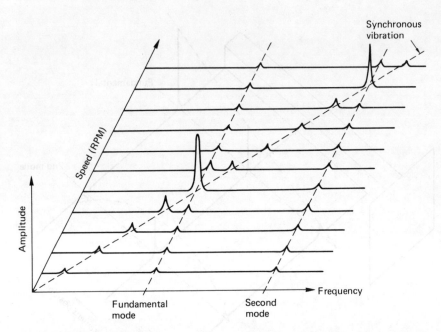

Figure 7.9 Waterfall diagram, formed by accelerating a rotating machine through a range of speeds and carrying out a spectral analysis at each speed. Notice that the largest responses (which may be dangerous) occur when the synchronous vibrations arising from out-of-balance forces coincide with a natural frequency.

that the phenomenon of shaft whirl occurs when the frequency of synchronous vibration (caused by out-of-balance forces) coincides with a natural frequency of the system. Waterfall diagrams are also used to detect instability in a rotating machine, which can arise because of internal friction or the dynamic behaviour of films of lubricating oil.

VECTOR RESPONSE DIAGRAMS

If the amplitude and phase of vibration are measured at a number of frequencies around a resonance, the results may be plotted as a series of vectors having a common origin as shown in figure 7.10. Such a diagram is known as a vector response diagram, or VRD. For systems with either hysteretic damping or low values of viscous damping the locus representing the ends of the vectors forms an arc of a circle passing through the origin. The resonant frequency f_n has a 90° phase angle, and the damping ratio for the resonant mode is (see figure 7.10):

$$\text{Damping ratio} = \frac{f_2 - f_1}{2f_n}$$

FREQUENCY DOMAIN ANALYSIS 169

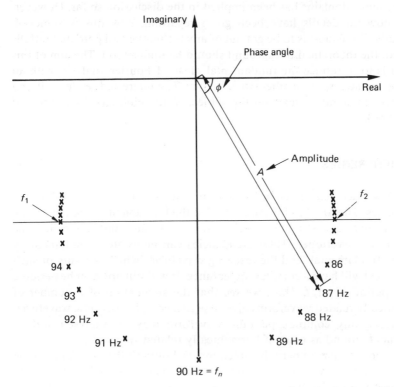

Figure 7.10 *Vector response diagram for a system with a resonance at 90 Hz. The vector shows the response of the system at 87 Hz, with phase angle ϕ and amplitude A.*

The amplitude and phase of a response signal can be determined directly by most modern spectrum analysers, some of which will also display a VRD. The advantage of this technique is that it allows the resonant frequency and damping to be determined separately for each mode of vibration in a system which is responding in several modes simultaneously.

Response tests on systems where several resonances are close together in frequency are difficult using amplitude measurements alone. It is possible to excite one mode and also have a significant contribution from adjacent modes, some being forced below and some above their resonant frequencies. In such a case the VRD consists of a number of arcs of circles which do not necessarily pass through the origin. However, provided a reasonable length of each arc is available for curve fitting, very fine resolution of the modal properties can be obtained by this method.

FOURIER ANALYSIS

The use of a transform to enable data to be inspected in either the time or the frequency domains has been implicit in the discussion so far. However, up to now no details have been given regarding how this is achieved. Generally, if the most is to be got out of any technique and potential pitfalls avoided, the theoretical background should be understood. The aim of this section is to introduce the fundamental ideas of Fourier analysis, without making any attempt at mathematical rigour. The interested reader will find references to standard texts on the subject in the bibliography at the end of the book.

FOURIER SERIES

It is well-known that a stretched string can resonate at a number of different frequencies. The lowest of these f is called the fundamental frequency. The higher resonances will have frequencies $2f$, $3f$ etc. and are called the harmonics or overtones. These frequencies can be excited simultaneously with the fundamental, and the resulting waveform (which is a sum of such resonances) will have a complex appearance. It will still however be periodic with a period $T = 1/f$. Thus, we see that the summation of a number of harmonically related waveforms gives rise to a complex periodic waveform. Conversely, any complex periodic waveform may be analysed and its constituents found as a set of harmonically related sinusoids.

Suppose we have a periodic function $f(t)$. Fourier's theorem for periodic functions states that any reasonable function having a period T may be represented in the form

$$f(t) = \frac{a_0}{2} + \sum_{n=1}^{\infty} \left\{ a_n \cos\left(\frac{2\pi n t}{T}\right) + b_n \sin\left(\frac{2\pi n t}{T}\right) \right\} \tag{7.1}$$

The term 'reasonable' means that the function $f(t)$ must satisfy certain conditions, known as Dirichlet's conditions, if the series expansion is to be valid. These conditions (which are met by almost all practical functions) can be stated as follows:

(a) The integral of $|f(t)|$ over one period must be finite.
(b) The number of jump discontinuities within an interval t must be finite.
(c) The number of maxima and minima of $f(t)$ must be finite within an interval t.

If $f(t)$ is periodic with period T then $f(t+nT) = f(t)$ (where $n = 0, \pm 1, \pm 2, \pm 3, \ldots$). Thus, if we take any interval between t_1 and t_2 such that $t_2 = t_1 + T$, the segment of the function between t_1 and t_2 will repeat itself indefinitely to the left and right of the interval as shown in figure 7.11(a).

Any periodic function f(t) is therefore completely known once its form is known within an interval between t_1 and $t_2 = t_1 + T$.

Alternatively, given a function f(t) which is known within an interval t_1 to $t_2 = t_1 + T$, we can construct an associated periodic function which is identical to f(t) within the interval t_1 to t_2, but which repeats itself indefinitely outside the interval as shown by figure 7.11(b). If $f_p(t)$ denotes this associated periodic function, a formal definition of $f_p(t)$ is

$$f_p(t) = f(t) \quad \text{where} \quad t_1 < t < t_2$$
$$f_p(t + np) = f(t) \quad n = \pm 1, \pm 2, \pm 3, \ldots$$

To obtain expressions for the coefficients a_n and b_n in the Fourier series is straightforward. If equation (7.1) is integrated, we obtain

$$\int_{t_1}^{t_2} f(t)\,dt = \frac{a_0}{2}\int_{t_1}^{t_2} dt + \sum_{n=1}^{\infty} a_n \int_{t_1}^{t_2} \cos\left(\frac{2\pi nt}{T}\right) dt$$
$$+ \sum_{n=1}^{\infty} b_n \int_{t_1}^{t_2} \sin\left(\frac{2\pi nt}{T}\right) dt$$

and since

$$\sum_{n=1}^{\infty} \int_{t_1}^{t_2} \cos\left(\frac{2\pi nt}{T}\right) dt = 0 \quad \text{and} \quad \sum_{n=1}^{\infty} \int_{t_1}^{t_2} \sin\left(\frac{2\pi nt}{T}\right) dt = 0$$

the expression for a_0 follows immediately:

$$a_0 = \frac{2}{T} \int_{t_1}^{t_2} f(t)\,dt \qquad (7.2)$$

To obtain expressions for a_n and b_n we shall need to use some important properties of sine and cosine, known as Fourier's integrals. These are listed below:

$$\int_{t_1}^{t_2} \cos\left(\frac{2\pi nt}{T}\right) \cos\left(\frac{2\pi mt}{T}\right) dt = 0 \quad (m \neq n)$$
$$= T/2 \quad (m = n) \qquad (7.3)$$

$$\int_{t_1}^{t_2} \sin\left(\frac{2\pi nt}{T}\right) \sin\left(\frac{2\pi mt}{T}\right) dt = 0 \quad (m \neq n)$$
$$= T/2 \quad (m = n) \qquad (7.4)$$

$$\int_{t_1}^{t_2} \sin\left(\frac{2\pi nt}{T}\right) \cos\left(\frac{2\pi mt}{T}\right) dt = 0 \quad (m \neq n)$$
$$= 0 \quad (m = n) \qquad (7.5)$$

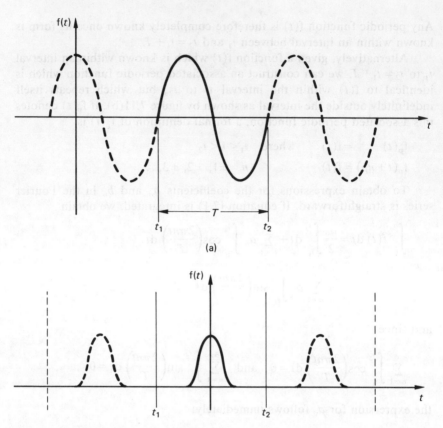

Figure 7.11 (a) A periodic function is completely defined once its form is known within an interval t_1 to $t_2 = t_1 + T$. (b) Associated periodic function produced by the Fourier series representation of a non-periodic function.

If both sides of equation (7.1) are multiplied by $\cos(2\pi mt/T)$ where $m = \pm 1, \pm 2, \pm 3, \ldots$, and the result integrated, we get

$$\int_{t_1}^{t_2} f(t) \cos\left(\frac{2\pi mt}{T}\right) dt = \frac{a_0}{2} \int_{t_1}^{t_2} \cos\left(\frac{2\pi mt}{T}\right) dt$$

$$+ \sum_{n=1}^{\infty} a_n \int_{t_1}^{t_2} \cos\left(\frac{2\pi nt}{T}\right) \cos\left(\frac{2\pi mt}{T}\right) dt$$

$$+ \sum_{n=1}^{\infty} b_n \int_{t_1}^{t_2} \sin\left(\frac{2\pi nt}{T}\right) \cos\left(\frac{2\pi mt}{T}\right) dt \quad (7.6)$$

Since sine and cosine are orthogonal functions all terms except those where $n = m$ are zero in the first summation, and all the terms in the final summation are zero. We saw earlier that the result of the first integral on the right-hand side of the equation was zero. Thus, the right-hand side of the equation reduces to $a_n(T/2)$, and replacing m by n we obtain the formula for the coefficient a_n as

$$a_n = \frac{2}{T}\int_{t_1}^{t_2} f(t) \cos\left(\frac{2\pi nt}{T}\right) dt \qquad (7.7)$$

Similarly, integration after multiplying by $\sin(2\pi mt/T)$ yields an expression for b_n:

$$b_n = \frac{2}{T}\int_{t_1}^{t_2} f(t) \sin\left(\frac{2\pi nt}{T}\right) dt \qquad (7.8)$$

To sum up, we have found expressions for the coefficients a_0, a_n and b_n and can thus express any periodic function $f(t)$ as a Fourier series by the use of equations (7.1), (7.2), (7.7) and (7.8). Even if $f(t)$ is not periodic, a Fourier series can still be used to represent $f(t)$ inside the interval t_1 to $t_2 = t_1 + T$. Outside the interval the series will produce an associated periodic function.

There would be little point in representing a function such as $f(t)$ by a Fourier series if an infinite number of terms of the series had to be evaluated. Fortunately however, it can be shown that a Fourier series is highly convergent, and a good approximation to $f(t)$ is usually obtained by the sum of only a few terms.

It was shown in figure 7.5(b) that a continuous, periodic signal always has a magnitude spectrum consisting of discrete lines. The height of the nth line of such a spectrum is given by $(a_n^2 + b_n^2)^{1/2}$. Its phase is given by $\arctan(b_n/a_n)$.

Even and Odd Functions

If $f(t)$ is an even or odd function the series expansion of $f(t)$ is simplified, with either the a_n or b_n coefficients becoming zero.

If $f(t)$ is an even function, that is $f(t) = f(-t)$, the integrand of equation (7.8) is odd. This is because the integrand is a product of the even function $f(t)$ and the odd function $\sin(2\pi nt/T)$. The product of an even function and an odd function is odd, and since the integral of an odd function over an interval symmetric about the origin is zero, the coefficients b_n are zero. An even function is accordingly represented by a Fourier cosine series:

$$f(t) = \frac{a_0}{2} + \sum_{n=1}^{\infty} a_n \cos\left(\frac{2\pi nt}{T}\right) \qquad (7.9)$$

If $f(t)$ is odd, that is $f(t) = -f(t)$, the integrands in expressions (7.2) and (7.7) for a_0 and a_n are both odd functions of t. Once again we are

174 INSTRUMENTATION FOR ENGINEERS

integrating an odd function over an interval which is symmetric about the origin, and a_0 and a_n are zero. An odd function may therefore be represented by the Fourier sine series:

$$f(t) = \sum_{n=1}^{\infty} b_n \sin\left(\frac{2\pi n t}{T}\right) \qquad (7.10)$$

It is often useful to be able to predict analytically the frequency content of a periodic signal. This is relatively straightforward for most common waveforms, as shown by the following examples.

Example 1

Figure 7.12 is a sketch of a triangular wave, which can be expressed as follows:

$$f(t) = 1 + \frac{4t}{T} \quad \text{for} \quad -\frac{T}{2} < t < 0$$

$$f(t) = 1 - \frac{4t}{T} \quad \text{for} \quad 0 < t < \frac{T}{2}$$

Since $f(t) = f(-t)$ the function is even, and the coefficients b_n will be 0. The coefficient a_0 is given by equation (7.2) as:

$$a_0 = \frac{2}{T} \int_{-T/2}^{+T/2} f(t)\, dt = \frac{2}{T} \int_{-T/2}^{0} \left(1 + \frac{4t}{T}\right) dt + \frac{2}{T} \int_{0}^{+T/2} \left(1 - \frac{4t}{T}\right) dt = 0$$

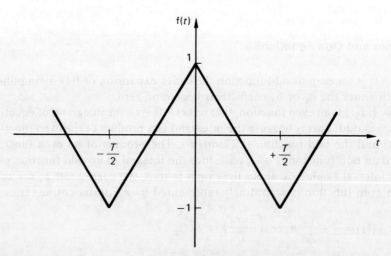

Figure 7.12 Triangular wave, defined by: $f(t) = 1 + 4t/T$ for $-T/2 < t < 0$, $f(t) = 1 - 4t/T$ for $0 < t < T/2$.

The coefficients a_n are found from equation (7.7):

$$a_n = \frac{2}{T} \int_{-T/2}^{+T/2} f(t) \cos\left(\frac{2\pi nt}{T}\right) dt$$

$$= \frac{2}{T} \int_{-T/2}^{+T/2} \cos\left(\frac{2\pi nt}{T}\right) dt + \frac{2}{T} \int_{-T/2}^{0} \frac{4t}{T} \cos\left(\frac{2\pi nt}{T}\right) dt$$

$$- \frac{2}{T} \int_{0}^{+T/2} \frac{4t}{T} \cos\left(\frac{2\pi nt}{T}\right) dt$$

$$= \frac{8}{T^2} \int_{T/2}^{0} (-t) \cos\left(\frac{2\pi nt}{T}\right) d(-t) - \frac{8}{T^2} \int_{0}^{T/2} t \cos\left(\frac{2\pi nt}{T}\right) dt$$

$$= -\frac{16}{T^2} \int_{0}^{T/2} t \cos\left(\frac{2\pi nt}{T}\right) dt$$

$$= -\frac{16}{T^2} \left\{ \left[t \frac{T}{2\pi n} \sin\left(\frac{2\pi nt}{T}\right) \right]_{0}^{T/2} - \frac{T}{2\pi n} \int_{0}^{T/2} \sin\left(\frac{2\pi nt}{T}\right) dt \right\}$$

$$= \frac{8}{T\pi n} \left[-\frac{T}{2\pi n} \cos\left(\frac{2\pi nt}{T}\right) \right]_{0}^{T/2}$$

$$= \frac{4}{\pi^2 n^2} (1 - \cos(n\pi))$$

that is

$$a_n = \frac{8}{\pi^2 n^2} \quad \text{when } n \text{ is odd,}$$

and

$$a_n = 0 \quad \text{when } n \text{ is even}$$

and the Fourier series representation of a triangular wave may then be obtained using equation (7.9).

Example 2

Figure 7.13 is a sketch of a square wave. With the choice of origin shown, f(t) is odd or antisymmetric. Thus, the Fourier expansion will be a sine series, and $a_n = 0$ for all n (including $n=0$).

The function f(t) is defined by the expressions:

$$f(t) = -1 \quad \text{for } -\frac{T}{2} < t < 0$$

$$f(t) = +1 \quad \text{for } 0 < t < \frac{T}{2}$$

176 INSTRUMENTATION FOR ENGINEERS

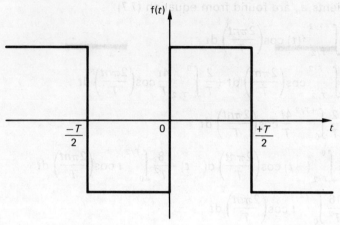

Figure 7.13 Square waves, defined by: $f(t) = -1$ for $-T/2 < t < 0$, $f(t) = +1$ for $0 < t < +T/2$.

and

$$f(t) = f(t+T)$$

Using equation (7.8), we have

$$b_n = \frac{2}{T} \int_{-T/2}^{+T/2} f(t) \sin\left(\frac{2\pi nt}{T}\right) dt$$

$$= \frac{2}{T} \left\{ \int_{-T/2}^{0} (-1) \sin\left(\frac{2\pi nt}{T}\right) dt + \int_{0}^{+T/2} (+1) \sin\left(\frac{2\pi nt}{T}\right) dt \right\}$$

$$= \frac{2}{T} \left\{ \left[\frac{T}{2\pi n} \cos\left(\frac{2\pi nt}{T}\right) \right]_{-T/2}^{0} - \left[\frac{T}{2\pi n} \cos\left(\frac{2\pi nt}{T}\right) \right]_{0}^{+T/2} \right\}$$

$$= \frac{1}{n\pi} \{ [1 - \cos(-n\pi)] + [1 - \cos(+n\pi)] \}$$

$$= \frac{2}{n\pi} (1 - \cos n\pi)$$

that is

when n is even, $b_n = 0$ and when n is odd $b_n = \frac{4}{n\pi}$

Notice that a different choice of origin would make $f(t)$ even. The reader is invited to prove that the function may then be represented by a Fourier cosine series where

$$a_0 = 0 \quad \text{and} \quad a_n = \frac{4}{n\pi} \sin\left(\frac{n\pi}{2}\right)$$

THE FOURIER TRANSFORM

A restriction on the use of the Fourier series is that it can only be applied to periodic signals. Many real signals are not periodic. For example, consider the response of a tuning fork when struck, the record of vibration experienced by a bridge as a vehicle crosses, or the voltage measured across a capacitor as it discharges.

Using the formulae

$$\cos x = \tfrac{1}{2}\{e^{jx} + e^{-jx}\}$$

$$\sin x = \frac{1}{2j}\{e^{jx} - e^{-jx}\}$$

the Fourier series can be rewritten:

$$f(t) = \sum_{n=-\infty}^{\infty} F_n \, e^{(j2\pi n t / T)} \qquad (7.11)$$

where $F_0 = a_0/2$ and $F_n = \tfrac{1}{2}(a_n + jb_n)$. Note that the coefficients a_n and b_n have been replaced by a set of coefficients F_n. By rearranging, an alternative more convenient expression for F_n can be derived:

$$F_n = \frac{1}{T} \int_{t_1}^{t_2} f(t) \, e^{(-j2\pi n t / T)} \, dt \qquad (7.12)$$

Equations (7.11) and (7.12) are known as the complex form of the Fourier series.

Suppose that the function $f(t)$ which we wish to represent by a Fourier series is a transient—that is, it is non-periodic. As discussed earlier, an associated periodic function can be constructed which is identical to $f(t)$ within the limits t_1 to $t_2 = t_1 + T$, but which repeats itself indefinitely outside this interval. However, if we want to represent our function over the whole range of t, it is necessary to let $t_1 \to -\infty$ and $t_2 \to +\infty$.

The effect of increasing the length of the interval T is to pack the frequency axis more densely and to reduce proportionately the magnitude of the coefficients F_n. At the limit when $T \to \infty$ the function $f(t)$ will be correctly represented for all values of t, the separation of the terms in the spectrum will be zero, and the coefficients F_n will be infinitesimally small. However, if we omit the factor $1/T$ from equation (7.12) they will remain finite as $T \to \infty$, and give the form of the distribution of amplitude. It is this which constitutes the Fourier transform of the function f(t). It is a continuous curve rather than a set of discrete coefficients. In other words, the Fourier Transform of a non-periodic function $x(t)$ is a continuous spectrum.

By a process similar to that described above a second transform (known as the reverse Fourier Transform) can be defined, which enables a time domain function to be reconstructed from frequency domain information.

The Fourier Transform is defined by the following pair of integrals:

$$S(f) = \int_{-\infty}^{\infty} x(t) e^{-j2\pi ft} \, dt \quad \text{(the forward transform)} \tag{7.13}$$

$$x(t) = \int_{-\infty}^{\infty} S(f) e^{+j2\pi ft} \, df \quad \text{(the reverse transform)} \tag{7.14}$$

The expression $e^{\pm j2\pi ft} = \cos 2\pi ft \pm j \sin 2\pi ft$ is known as the kernel of the Fourier transform. $S(f)$ is known as the Fourier transform of $x(t)$. $S(f)$ is in general complex, and contains amplitude and phase information for all the frequencies which make up $x(t)$ even though $x(t)$ is not periodic.

Since $S(f)$ is in general complex, if we want to plot $S(f)$ as a function of frequency we either have to plot two graphs (one representing the real, one the imaginary parts of $S(f)$), or we have to resort to some three-dimensional representation.

$S(f)$ can be written:

$$S(f) = a(f) - jb(f) \tag{7.15}$$

where $a(f)$ is the real and $-b(f)$ the imaginary part of $S(f)$. If $x(t)$ is an even function, then by definition $x(t) \sin 2\pi ft$ is odd. Hence

$$b(f) = \int_{-\infty}^{\infty} x(t) \sin 2\pi ft \, dt = 0 \tag{7.16}$$

Thus, in the case of an even function the complex transform $S(f)$ degenerates to a pure real function with no imaginary part.

Similarly, it can be shown that if $x(t)$ is odd, $a(f) = 0$ and for an odd function $S(f)$ is pure imaginary.

These results are of great utility in evaluating the Fourier Transforms of various common functions. As an example, let us consider the Fourier Transform of a tone burst, or cosine wave of short duration. This function arises in practice every time you switch on an oscillator and some time later switch it off again. In the time domain, it might have the form shown in figure 7.14. The Fourier Transform of $x(t)$ in this case is

$$S(P) = \int_{-\infty}^{\infty} x(t) e^{-j\omega t} \, dt = \int_{-T/2}^{+T/2} \cos Pt \, e^{-j\omega t} \, dt$$

(where ω = angular frequency = $2\pi f$).

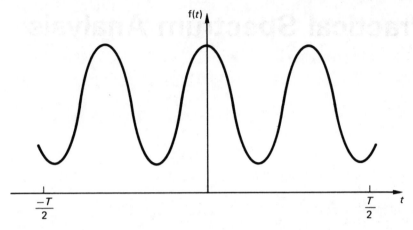

Figure 7.14 Cosine tone burst, defined by: $f(t) = \cos(Pt)$ *for* $-T/2 < t < T/2$, $f(t) = 0$ *elsewhere.*

This is a standard integral which can be carried out analytically, although the process is somewhat lengthy. However, recalling the properties of odd and even functions discussed above, we can take a short cut, since $x(t)$ in this case is obviously even. Therefore, its Fourier Transform is a pure real function—that is, it is equal to $a(\omega)$. Hence

$$S(\omega) = a(\omega) = \int_{-\infty}^{\infty} x(t) \cos \omega t \, dt$$

$$= \int_{-T/2}^{+T/2} \cos Pt \cos \omega t \, dt$$

$$= \tfrac{1}{2} \int_{-T/2}^{+T/2} \{\cos(\omega - P)t + \cos(\omega + P)t\} \, dt$$

$$= \frac{T}{2} \left\{ \frac{\sin \tfrac{1}{2}(\omega - P)T}{\tfrac{1}{2}(\omega - P)T} + \frac{\sin \tfrac{1}{2}(\omega + P)T}{\tfrac{1}{2}(\omega + P)T} \right\}$$

The number of cycles of the waveform in the pulse, that is the number of cycles of angular frequency P occurring in time T, is given by $PT/2\pi$. If there are a large number of cycles in the pulse, that is if PT is large, then the second term in the expression for $S(\omega)$ will be small for the positive range of ω. The curve for $S(\omega)$ then becomes the curve $\sin x/x$ centred at $\omega = P$.

When however there are only a small number of cycles in the waveform, either because P is small (that is, a low frequency), or because T is small (that is, the oscillator was only switched on for a short time), then the second term makes an important contribution to $S(\omega)$ and the curve changes its shape.

Chapter 8
Practical Spectrum Analysis

INTRODUCTION

In the previous chapter we examined the theoretical background to the Fourier series representation of periodic waveforms, and the use of the Fourier Transform to examine data in the frequency domain. In this chapter we shall concentrate on the practical side of spectral analysis. As with almost all analysis techniques, there are a number of pitfalls awaiting the unwary which can lead to erroneous results.

The instruments used for studying frequency domain behaviour are called spectrum analysers. They come in several types, but are usually divided into two classes according to whether their internal functioning is analogue or digital. Nowadays, most spectrum analysers are digital. However, analogue analysers are still common, and the following brief descriptions outline the operating principles of the different types of analogue analyser.

ANALOGUE ANALYSERS
Discrete Stepped Filter Analysers

Figure 8.1 shows a block diagram of a typical fixed filter analyser. The signal is applied in parallel to a bank of filters, contiguous in the frequency domain, which together cover the frequency range of interest. A detector is connected sequentially to the various filter outputs, and thus successively measures the output power within each frequency band. Note that when using this instrument it is not necessary to wait for the filter response time

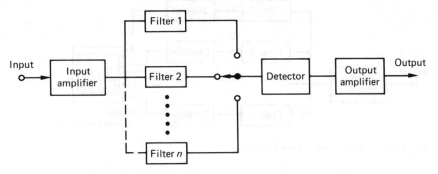

Figure 8.1   Discrete stepped filter analyser.

before making a measurement, but only for that of the detector, which is usually very fast.

This approach is normally only used for bandwidths down to $\frac{1}{3}$ of an octave, since the number of filters needed to obtain higher resolution is prohibitive.

Swept-Filter Analysers

For narrow-band analysis it is more common to use a single filter with a tunable centre frequency, as illustrated by the block diagram of figure 8.2. The discrete-filter analyser described earlier provides a number of spectral power estimates at discrete frequencies, which correspond to the centre frequencies of the filters used. The spectrum obtained from a swept-filter analyser however is continuous. Each point on the spectral curve represents an integration of the true spectrum over a frequency range corresponding to the bandwidth of the filter.

Parallel Analysers (Real Time)

The two types of analyser just discussed are known as sequential or serial analysers, since the analysis is carried out at each frequency in turn. Thus, the implicit assumption is made that the signal is *stationary*. This means

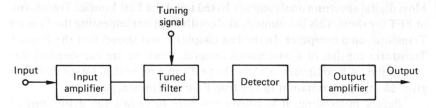

Figure 8.2   Sweeping filter analyser.

182 INSTRUMENTATION FOR ENGINEERS

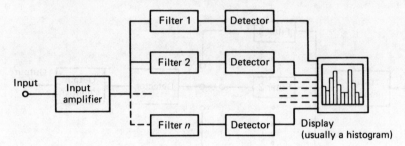

Figure 8.3   Real-time parallel analyser.

that the average value of the signal is a constant, and that the average of the product of two samples of the signal taken at separate times (the autocorrelation) is also a constant which depends only on the separation between the two samples. Another way of determining whether a signal is stationary or not is to see whether its spectrum changes with time. If the signal is not stationary, the form of the corresponding spectrum will probably change between the first and last filtering operations. Sometimes a signal can be forced to be stationary, by recording a section of it on a tape loop which is played back repetitively. However, this procedure can give misleading results, since it imposes an artificial periodicity on the time domain data.

So-called real-time analysers such as that shown in figure 8.3 obtain the whole spectrum in parallel almost instantaneously, from the same section of signal. Thus, these instruments can not only display the spectral content of non-stationary signals, but are much faster in operation than sequential analysers. The most direct way of implementing a real-time analyser is to apply the signal to a parallel bank of filter/detector channels as shown in figure 8.3. The speed with which the results are generated makes it desirable to view them on a continuously updated screen. The resolution of this type of analyser is usually fairly low, because of the cost and complexity involved in constructing large numbers of narrow-bandwidth filters.

DIGITAL ANALYSERS

Most digital spectrum analysers are based upon the Fast Fourier Transform, or FFT for short. This is a numerical algorithm for implementing the Fourier Transform on a computer. In the last chapter it was shown that the Fourier Transform consists of a continuous integral. Since we are carrying out this integration on a computer, we have to perform a numerical integration that gives us an approximation to the true Fourier Transform.

Before proceeding, it is convenient here to restate the definitions of the Fourier series and the Fourier Transform.

PRACTICAL SPECTRUM ANALYSIS 183

Any periodic time function can be represented as an infinite sum of weighted sine and cosine functions of the proper frequencies. This allows time domain functions to be interpreted by analysis of their frequency content. This can be expressed mathematically as

$$x(t) = \frac{a_0}{2} + \sum_{n=1}^{\infty} \left\{ a_n \cos\left(\frac{2\pi n t}{T}\right) + b_n \sin\left(\frac{2\pi n t}{T}\right) \right\} \quad (8.1)$$

where T is the period of the time function $x(t)$, that is

$$x(t) = x(t+T)$$

and the coefficients a_0, a_n and b_n are

$$a_0 = \frac{2}{T} \int_{t_1}^{t_2} x(t) \, dt \quad (8.2)$$

$$a_n = \frac{2}{T} \int_{t_1}^{t_2} x(t) \cos\left(\frac{2\pi n t}{T}\right) dt \quad (8.3)$$

$$b_n = \frac{2}{T} \int_{t_1}^{t_2} x(t) \sin\left(\frac{2\pi n t}{T}\right) dt \quad (8.4)$$

where t_1 and $t_2 = t_1 + T$ are the end points of the interval T. The above is known as Fourier's theorem.

A restriction on the use of Fourier series is that the time signal must be periodic. This restriction may be overcome if we allow the period T of the signal to approach infinity. The resulting evaluation of the Fourier series is known as the Fourier Transform, and it is defined by the following pair of integrals:

$$S(f) = \int_{-\infty}^{\infty} x(t) \, e^{-j2\pi f t} \, dt \quad \text{(the Forward Transform)} \quad (8.5)$$

$$x(t) = \int_{-\infty}^{\infty} S(f) \, e^{+j2\pi f t} \, df \quad \text{(the Reverse Transform)} \quad (8.6)$$

The expression $e^{\pm j2\pi f t} = \cos 2\pi f t \pm j \sin 2\pi f t$ is known as the kernel of the Fourier Transform. $S(f)$ is known as the Fourier Transform of $x(t)$. It contains amplitude and phase information for all the frequencies which make up $x(t)$, and $x(t)$ need not be periodic.

To obtain the Fourier Transform of a time domain signal on a digital system such as a computer, the continuous time signal must be represented by a set of discrete data points—or to use the jargon, the signal must be sampled and digitised. However, the Fourier Transform integral is continuous, or in other words it implies that the time interval between the data points is infinitesimal, that is

$$\Delta t \to dt$$

This is obviously impossible in a practical system. To get around the problem, we replace the Fourier Transform integral with a summation:

$$S'(f) = \Delta t \sum_{n=-\infty}^{\infty} x(n\Delta t)\, e^{-j2\pi f n \Delta t} \tag{8.7}$$

where $x(n\Delta t)$ is the value of the signal measured at the sampling intervals Δt. The summation shows that we can still calculate a valid Fourier Transform, even though we are dealing with a discrete sample with 'real' sampling intervals Δt, large compared with the infinitesimal dt.

However, the magnitude and phase information for all the frequencies contained in $S(f)$ is not accurate when a Fourier Transform is calculated in this manner. The function $S'(f)$ accurately describes the spectrum of $x(t)$ only up to a frequency F_{max}, dependent on the sampling interval as discussed later (see Shannon's Sampling Theorem).

Examination of equation (8.7) shows that if $S'(f)$ is to be calculated accurately, an infinite number of samples of the input waveform will be required. Again, this is obviously impossible. Any practical spectrum analyser has to deal with a finite number of data points accumulated over a finite length of time. If the data acquisition time is T, the number of data points N, and the sampling interval Δt, then

$$N = T/\Delta t$$

Restricting the sampling time to T is equivalent to truncating the summation, and we cannot therefore calculate magnitude and phase values for an unlimited number of frequencies between 0 and F_{max}. The truncated version of the summation will not produce a continuous frequency spectrum. We can express this truncated summation as

$$S''(m\Delta f) = \Delta t \sum_{n=0}^{N-1} x(n\Delta t)\, e^{-j2\pi m \Delta f n \Delta t}$$
$$(\text{for } m = 0, 1, 2, \ldots, N-1) \tag{8.8}$$

This expression is known as the Discrete Fourier Transform or DFT. It is a sampled Fourier series with N real-valued time domain data points, and it can easily be implemented on a computer. A common algorithm for evaluating the DFT is the Fast Fourier Transform, or FFT. This algorithm, or one of its variants, is the basis of most digital spectrum analysers.

THE FAST FOURIER TRANSFORM

The Fast Fourier Transform (FFT) is an algorithm for efficiently obtaining the Direct Fourier Transform (DFT) of a function, with a smaller number of arithmetic operations than are required with the direct approach. It was first devised in 1965 by Cooley and Tukey, and has revolutionised signal analysis. It is without doubt one of the most important analysis techniques available to the instrumentation engineer.

The algorithm was at first limited to applications in which it could be implemented on a large computer. However, the advent of the microprocessor has meant that relatively low-cost stand-alone analysers are now available. These are now the most common form of frequency analyser, and have the advantage that they are equally suitable for working in the frequency or modal domains.

The inner workings of the FFT algorithm are complicated and we shall not go into details here. It is sufficient to know that the algorithm operates on an array of N complex data points in one domain, and produces an array of $N/2$ data points in the other domain. Both input and output data occupy the same array, with the second half of the output data array containing information which is normally redundant. The FFT algorithm has one very important implication for the user:

the number of data points must be a power of 2

In other words, the number of input samples must be

$$512 = 2^9, \quad 1024 = 2^{10}, \quad 2048 = 2^{11}, \quad \text{and so on}$$

The largest number of points the analyser can handle is determined by the memory available. Commonly, an analyser will store and transform $1024 = 1\,\text{K}$, $2\,\text{K}$, $4\,\text{K}$ or $8\,\text{K}$ at a time. Some systems have a dual memory, so that one block of data is being transformed while a further block is being acquired.

Obviously, the larger the capacity of the transform, the greater the available resolution. Usually it is not worth carrying out an FFT on less than 512 points, since the transformed data will be too widely spaced (that is, of very low resolution).

We saw earlier that one of the consequences of the DFT summation is that if N real data points are input, $N/2$ (complex) data points are output. So far, we have not made any real distinction between the time samples $x(t)$ and the frequency spectrum values $S(f)$. In the most common practical situation, the $x(t)$ values will be real and $S(f)$ will be complex. The FFT algorithm is equally valid for real or complex data, and strictly should be thought of as transforming complex data in one domain into complex data in the other domain. When used on real time domain data however, there will be an important redundancy. The imaginary part of each input data value will be zero, and thus half the memory will be used for storing zeros, which is wasteful. It is possible to remove this redundancy by means of a modification to the algorithm, so that N real values are transformed as though they were $N/2$ complex values, and the result manipulated to give the correct spectrum. Most FFT analysers operate in this fashion.

The output array from an FFT algorithm for N real-valued samples is normally arranged as shown below:

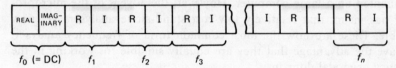

As shown above, the frequency range is always from DC (zero) to the Nyquist frequency f_n (=half the sample rate). This is independent of the number of samples in the record. (From table 8.1 it can be seen that increasing the number of samples increases the resolution, NOT the maximum frequency in the spectrum.)

We saw earlier that it is necessary to use low-pass filters to guard against aliasing, and since the spectrum is calculated up to f_n, it is desirable to place the cut-off of the filter as close as possible to this. However, since it is impossible to construct an ideal low-pass filter, the cut-off usually has to be placed at a frequency rather lower than f_n. For this reason the B&K 2031 FFT analyser, for example, calculates a 512-point spectrum from 1 K of input samples, but only displays the first 400 points as valid data.

Practical Use of the FFT for Spectral Analysis

Let a continuous analogue signal from a sensor be sampled at regular intervals Δt, as shown in figure 8.4. If N samples are taken, the length of the data record must be

$$T = N\Delta t \qquad (8.9)$$

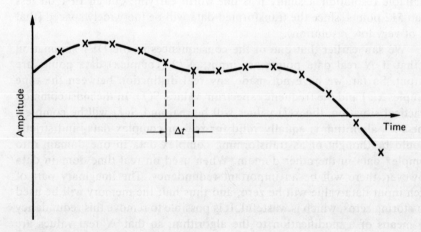

Figure 8.4 Sampled time domain data.

Table 8.1 Choice of Sampling Parameters

Choose a convenient value for parameter shown	Chosen parameter fixes (because)	Make either of remaining two (not both) as convenient as possible by choosing N
Δt	F_{max} ($F_{max} = 1/2\Delta t$)	$T\ (=N\Delta t)$ $\Delta f\ (=1/N\Delta t)$
T	Δf ($\Delta f = 1/T$)	$\Delta t\ (=T/N)$ $F_{max}\ (=N\Delta f/2)$
F_{max}	Δt ($\Delta t = 1/2F_{max}$)	$T\ (=N\Delta t)$ $\Delta f\ (=1/N\Delta t)$
Δf	T ($T = 1/\Delta f$)	$\Delta t\ (=T/N)$ $F_{max}\ (=N\Delta f/2)$

The data shown in figure 8.4 is in the time domain. It can be described in the frequency domain by means of a Fourier Transform, which will produce a set of frequency domain points defined by the following parameters:

Δf = frequency interval between points

$N/2$ = number of frequency domain points

F_{max} = maximum frequency = $(N\Delta f/2)$

Note that in general, the following relationships between time and frequency domain data hold:

$\Delta f = 1/T$ \hfill (8.10)

$\Delta t = 1/(2N\Delta f) = 1/(2F_{max})$ \hfill (8.11)

It is important that the right choices are made when deciding how to digitise an analogue signal. This is particularly vital if no analogue record (such as an FM tape recording) is being made of the signal—a wrong decision at this stage usually means a wasted experiment!

For a particular application, table 8.1 can be used to help. N (the number of points in the time domain) is always a power of 2 when using the FFT. N is normally fixed by the capabilities of the instrument you are using.

Example Calculation

Suppose we want to examine the spectrum of noise from a gearbox. We wish to look at the frequency range from DC to 2 kHz, with a resolution of 4 Hz. How many data points should we take?

We want $\Delta f = 4$ Hz. Therefore $T = 1/\Delta f = 0.25$ s

$F_{max} = 2000 = N\Delta f/2$

If $\Delta f = 4$ Hz, N must be 1000 points

In an FFT analyser we should therefore use 1024 points (2^{10}), which would give us a slightly better resolution than we require.

ALIASING AND SHANNON'S SAMPLING THEOREM (SST)

The phenomenon of aliasing can cause errors in measurements made by an instrumentation system which samples a continuous signal at regular intervals. This effect is best demonstrated by looking at an example. Consider the temperature monitoring device shown in figure 8.5. The system is set up to print out the temperature in a room once a second. As the temperature of a room only changes slowly, we would expect each reading to be almost the same as the last, as shown by figure 8.6. We are therefore sampling faster than is necessary in this case.

On the other hand, suppose the same arrangement is used to monitor the temperature of part of an engine, where the temperatures fluctuate more rapidly. If the temperature were to cycle once every two seconds, the result would indicate that the temperature was constant as shown in figure 8.7. We have not sampled fast enough to see the temperature fluctuations.

If we sample slower than twice the frequency of the input signal, a false low-frequency temperature variation appears, as shown by figure 8.8. This is the phenomenon of aliasing. When the sample rate is less than or equal to twice the highest frequency present in the input voltage, an *alias frequency* appears. This is expressed by *Shannon's Sampling Theorem*, which states that a sampled time signal must not contain components at frequencies

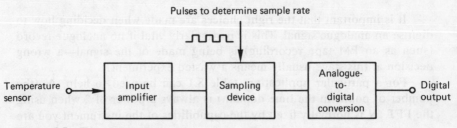

Figure 8.5 Temperature monitoring system.

PRACTICAL SPECTRUM ANALYSIS 189

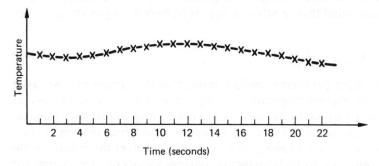

Figure 8.6 The temperature in a room, sampled at 1 Hz.

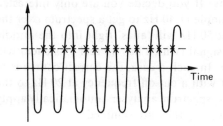

Figure 8.7 Sample rate 2 × signal frequency.

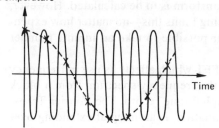

Figure 8.8 Sample rate less than 2 × frequency.

above half the sampling rate (the so-called Nyquist frequency). If the frequency of the input voltage f_{in} is greater than half the sample rate f_s, the spurious signal that results will appear to have a frequency f_{alias} given by

$$f_{alias} = f_s - f_{in} \tag{8.12}$$

The requirement for correct measurement, that the sample rate be faster than twice the highest frequency contained in the input signal, is known as the *Nyquist Criterion*. In the case of the room temperature measurement, we can be reasonably sure of the maximum rate at which the temperature might change. What we cannot be so sure of is whether the outside world will spoil things for us by introducing spurious signals into the system, for instance from mains wiring or from nearby radio transmitters. The only way to be certain that the input frequency range is limited is to add a low-pass filter before the sampler and the ADC. Such a filter is often called an antialias filter.

A common mistake is to misinterpret Shannon's sampling theorem as meaning that you have to sample at more than twice the highest frequency you are interested in. This is wrong! If you decide you are only interested in frequencies up to 15 Hz, and sample at 40 Hz to get a spectrum over the range DC to 20 Hz, mains hum at 50 Hz will 'alias' right into the middle of your spectrum and an aliasing signal (which has no real existence) will be seen with a frequency of 10 Hz. In such a case, the correct course is to apply a low-pass antialiasing filter with a cut-off frequency at 20 Hz to the signal before sampling occurs. Many spectrum analysers automatically apply antialias filtering to input signals before they are sampled.

WINDOWING

In the previous chapter we saw how truncation of a cosine tone burst affects its spectrum. To obtain a sharp peak in the magnitude spectrum, it is necessary to include as many cycles of the periodic function as possible in the record from which the Fourier Transform is to be calculated. However, the instrumentation system you are using limits this—no matter how expensive the spectrum analyser, it is never possible to take an infinite number of samples!

There is another property of the FFT which affects its use in frequency domain analysis. The FFT computes the frequency spectrum from a block of samples of the input, called a sampled time record. The Fourier Transform, to which the FFT is an approximation, is an integral extending over all time. Thus, the implicit assumption is made that the record contained in the sampled block is repeated throughout time.

PRACTICAL SPECTRUM ANALYSIS 191

This does not cause a problem when we are dealing with a transient such as that shown in figure 8.9, since the 'join' between the repeats is smooth and will not distort the spectrum.

However, consider the case of a continuous signal such as a sine wave. If the time record is arranged (usually by a judicious choice of the sample rate) so that it contains an integral number of cycles, then the assumption of continuity exactly matches the actual input waveform, as shown in figure 8.10. In such a case, the input waveform is said to be periodic in the time record. The magnitude spectrum obtained in such a case is shown in figure 8.11.

Figure 8.12 demonstrates the effect that occurs when an input is not periodic within the time record. The FFT is computed on the basis of the highly distorted waveform shown. We saw in chapter 7 that the (magnitude) spectrum of a continuous sine wave consists of a single line. The spectrum of the function shown in figure 8.12 will be very different. As a general rule, functions that are sharp or spiky in one domain appear spread-out in the

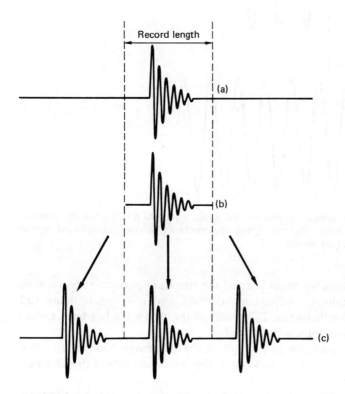

Figure 8.9 Smooth joining of the assumed repeats of a transient: (a) the signal; (b) truncated record; (c) assumed repeats.

192 INSTRUMENTATION FOR ENGINEERS

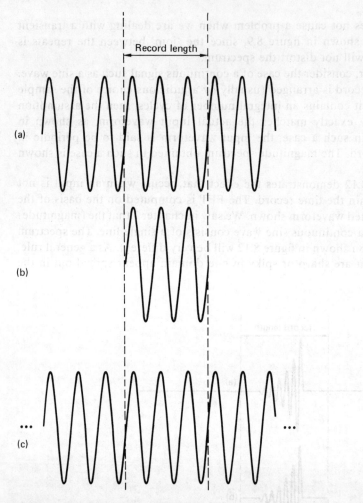

Figure 8.10 *When a signal is periodic within the truncated time record, the assumed repeats match the input: (a) the signal; (b) truncated record; (c) assumed repeats exactly match the input signal.*

other domain. Thus, we should expect the spectrum of our sine wave to be spread out throughout the frequency domain, and as shown in figure 8.13 this is exactly what happens. The power of the sine wave has been spread throughout the spectrum as predicted.

This smearing of energy throughout the frequency domain is known as leakage. Energy is said to leak out of one resolution line of the FFT into all the other lines.

It is important to realise that leakage occurs because we have taken a finite time record. For a sine wave to have a line spectrum, it would have

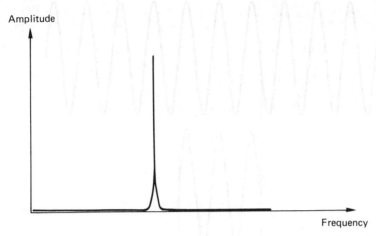

Figure 8.11 Magnitude spectrum obtained from sinusoid periodic within the time record.

to exist for all time. If we had an infinite time record, the FFT would calculate the line spectrum exactly. However, since we are not willing to wait forever to measure a spectrum, we have to put up with a finite record. This can cause leakage, unless we can arrange for our signal to be periodic in the time domain.

It is obvious from figure 8.13 that the problem of leakage is severe enough to mask small signals, if they are close in frequency to the main sine wave. A partial solution to this problem is provided by a technique known as windowing.

If we return to the problem of a sine wave that is not periodic in the time domain (figure 8.12), we see that the difficulty seems to be associated with the ends of the time record. The central part is a good sine wave. If the FFT can be made to ignore the ends and concentrate on the middle, we would expect to get closer to the correct single line spectrum in the frequency domain.

If we multiply our time record by a function that is zero at the ends of the time record, and large in the middle, we shall concentrate the FFT on the centre of the record. One such function is shown in figure 8.14(c). Such functions are called window functions, since they force the analyser to look at time domain data through a narrow window.

Figure 8.15 shows the vast improvement obtained by windowing data that is not periodic in the time domain. However, it is important to realise that by windowing it we have tampered with the data, and that we cannot therefore expect perfect results. The FFT now assumes that the data looks like figure 8.14(d). This has a magnitude spectrum which is closer to the correct single line, but it is still not exact. Figure 8.16 demonstrates that

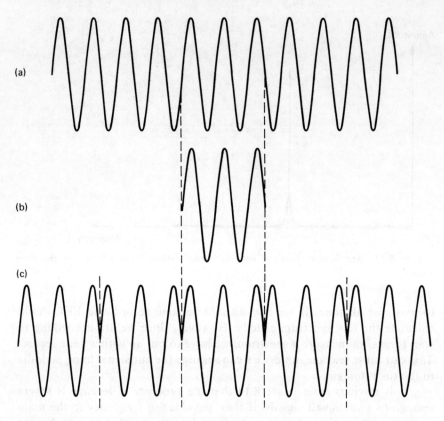

Figure 8.12 Input signal not periodic within time record, and resulting assumed input is distorted: (a) the signal; (b) truncated record; (c) assumed repeats do not match the input signal.

the windowed data does not have as narrow a spectrum as unwindowed data which is periodic in the time domain.

A number of functions can be used to window data, but one of the most common is known as the Hanning or $\cos^2$ window. This tails-off the data at both ends in a $\cos^2$ shape, and is illustrated in figure 8.17. The Hanning window is also commonly used when measuring continuous stationary random noise.

The Hanning window does a good job on continuous signals such as sinusoids and random noise. However, it is unsuitable for use when calculating the spectrum of a transient. A typical transient is shown in figure 8.18(a). If it is multiplied (windowed) by the Hanning window shown in figure 8.18(b), we should get the highly distorted signal shown in figure 8.18(c). If the transient represents, for example, the record of vibration of a bell after it is struck, all the vibration in higher modes occurs in the early part

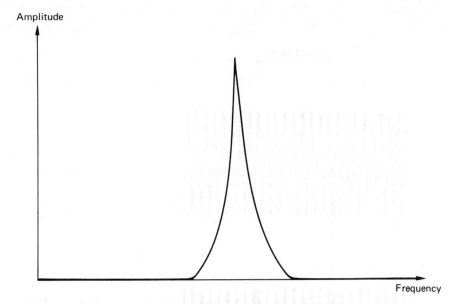

Figure 8.13 Distorted spectrum resulting from signal of figure 8.12.

of the record and is likely to be lost after windowing. The spectrum of the transient, with and without a Hanning window, is shown in figure 8.19. This Hanning window has taken the transient, which naturally has energy spread widely through the frequency domain, and has distorted it until it resembles a sine wave. For transients, using a Hanning window is dangerous and should be avoided.

Many transients have the property that they are zero at the beginning and end of the time record. Recalling that we introduced windowing in the case of continuous signals to force the time record to be zero at the beginning and end of the data, we see that for this kind of record there is no need to window the data. Functions like this which do not need a window are known as self-windowing. Such functions generate no leakage in the FFT. We can consider the data as being viewed through a rectangular or uniform window, created by our action of turning the recording instrument on and off.

CHOICE OF WINDOW

The tone burst discussed in chapter 7 can be considered as an infinite sinusoid multiplied by a rectangular window, as shown in figure 8.20. The operation of multiplication in the time domain corresponds to convolution

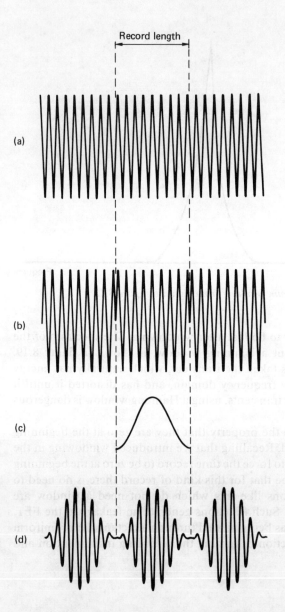

Figure 8.14 The effect of windowing a continuous signal: (a) the signal; (b) assumed repeat; (c) window function; (d) assumed repeat, windowed data.

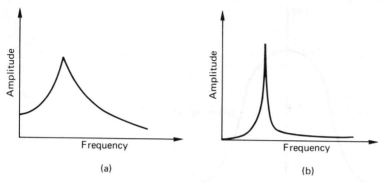

Figure 8.15 Spectra obtained from a sinusoid not periodic within the record: (a) without window; (b) with Hanning window function.

of the respective spectra in the frequency domain. Changing the length of the rectangular window changes its spectrum, and consequently altering the length of the tone burst alters the resulting spectrum, as was demonstrated analytically in chapter 7.

It can be seen from the discussion above that the influence of a given window function can best be assessed by examining it in the frequency domain. Figure 8.21 shows a comparison between the spectra of four commonly used windowing functions. The functions shown are a rectangular window, Hanning (cosine^2), Hamming (Hanning on a small rectangular pedestal) and a Gaussian function. The comparison is made with all windows having the same length. In the case of the Gaussian function (which

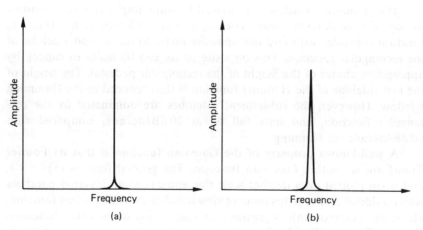

Figure 8.16 Windowed data, and unwindowed data which is periodic within the time record: (a) periodic within record—no leakage; (b) non-periodic within record—Hanning window.

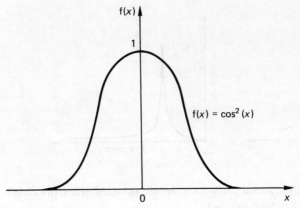

Figure 8.17 *The Hanning or cosine^2 window.*

is theoretically infinitely long) the window length is taken to be 7 times the standard deviation, which means that the effect of truncating the window will not be seen unless the dynamic range of the signal is greater than 60 dB.

Table 8.2 compares the window functions of figure 8.21, and gives numerical values for the 3 dB bandwidth, the height of the largest sidelobe, and the rate of fall-off of the sidelobes.

From table 8.2 and figure 8.21 it can be seen that the Hanning window gives a much better performance than the rectangular function. A Hanning window is generated by multiplying the time domain data by a cosine^2 function, which has a maximum value of 1 and is zero at both ends. Because of this ease of generation the Hanning window is available on almost all analysers.

The Hamming window is produced by mounting a Hanning window on top of a small rectangular pedestal. The first sidelobe of the Hanning function coincides with and has opposite phase to the second sidelobe of the rectangular function. This opposite phase can be made to cancel, by appropriate choice of the height of the rectangular pedestal. The height of the first sidelobe of the Hanning function is thus reduced in the Hamming window. However, the subsequent sidelobes are dominated by the rectangular function, and only fall off at 20 dB/decade, compared with 60 dB/decade for Hanning.

A well-known property of the Gaussian function is that its Fourier Transform is another Gaussian function. The general form is $\exp(-x^2)$, and when plotted on a decibel scale this appears as an inverted parabola with no sidelobes. From this point of view it makes an ideal window function. However, its bandwidth is greater than that of any of the other functions shown. Where the length of a record is restricted because of equipment or other limitations, probably a Hanning or Hamming window is the best choice.

PRACTICAL SPECTRUM ANALYSIS 199

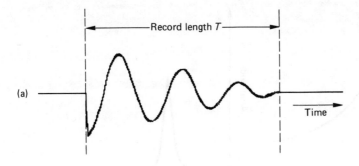

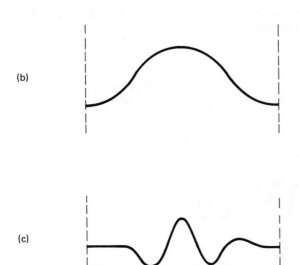

Figure 8.18 The effect of windowing a transient signal: (a) a typical transient; (b) window function; (c) windowed transient.

To conclude this chapter, the following glossary covers many of the terms commonly associated with spectrum analysis.

GLOSSARY OF FFT ANALYSER TERMINOLOGY

Frequency range: As described above, this is always from zero (DC) to the Nyquist frequency (that is, half of the sampling frequency).

200 INSTRUMENTATION FOR ENGINEERS

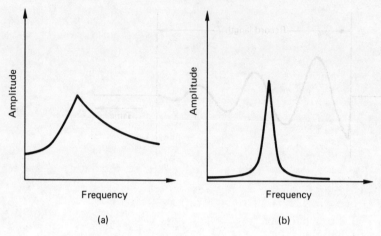

Figure 8.19 Spectrum of transient from figure 8.18, with and without Hanning window: (a) unwindowed; (b) Hanning window.

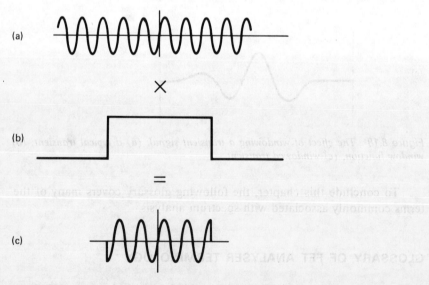

Figure 8.20 Tone burst as an infinite sinusoid multiplied by a rectangular window: (a) infinite sinusoid; (b) rectangular window; (c) tone burst.

PRACTICAL SPECTRUM ANALYSIS

Table 8.2 Comparison of Window Functions

Name	3 dB bandwidth	Size of largest sidelobe	Sidelobe fall-off rate
Rectangular	$0.9/T$	-13 dB	20 dB/decade
Hanning	$1.4/T$	-32 dB	60 dB/decade
Hamming	$1.3/T$	-42 dB	20 dB/decade
Gaussian	$1.8/T$	None	No sidelobes

Resolution: This is defined as the frequency interval between adjacent lines in the spectrum. It can be calculated from the number of samples in the time record:

$$R = f_s/N = 2f_n/N \quad \text{(where } f_s = \text{sample rate; } f_n = f_s/2\text{)}$$

Bandwidth: The term originates from the use of bandpass filters, which pass only that part of the total power whose frequency lies within a finite range (the bandwidth). The concept can be understood by considering an ideal filter, which transmits at full power all signals lying within its passband and completely attenuates all signals at other frequencies. The FFT can be thought of as a bank of ideal filters (one for each point in the spectrum), each having a passband or bandwidth determined by the sample rate and record size.

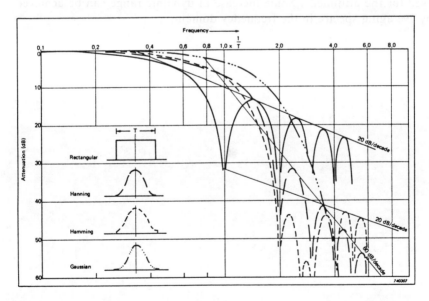

Figure 8.21 Comparison of the spectra of some common time window functions (reproduced by courtesy of Brüel and Kjaer).

The bandwidth is determined by the resolution of the analyser, and by any time window applied to the data. For linearly weighted data the bandwidth is equal to the resolution, that is

$B_{\text{eff}} = R$

If any time window is applied to the data then the bandwidth of the result is equal to the bandwidth of the weighting function. In particular the common Hanning window results in an effective noise bandwidth:

$B_{\text{eff}} = 1.5R$

One means of obtaining a finer resolution is to add zeros to the data record, but this is of course at the expense of a larger transform size and time. For example, if a record of length N is extended to $2N$ with zeros and a transform of size $2N$ performed, then the resolution will be $f_s/2N$ while the bandwidth will still be f_s/N. Another way of thinking about this is to imagine that we have applied a rectangular weighting function of length N to a data record of length $2N$.

Dynamic range: This is determined by the number of bits with which the input data is represented, and is defined as for an analogue-to-digital converter. As a rough rule-of-thumb, 6 dB of dynamic range are obtained for every bit of the input data, and thus 72 dB is obtained from analogue-to-digital conversion with 12 bits. The dynamic range is generally unaffected by the FFT calculation if four extra bits (that is, 16 bits for 12-bit data) are used for the arithmetic. Some increase in dynamic range can be achieved by averaging spectra in the frequency domain.

Chapter 9
Correlation and Spectral Analysis

INTRODUCTION

It is routine nowadays for signal analysis to be carried out in the frequency domain. However, the decision to apply a Fourier Transform to the data is often taken without much thought. Consideration should first be given to alternative methods. There are a number of excellent analysis techniques available for investigating the characteristics of time domain data. In this concluding chapter we shall introduce correlation, which is probably the best-known time domain analysis technique. There are many others, but a detailed discussion of them all would require more space than is available in this book. The reader who needs more than an introduction to the subject should look at some of the excellent books available, references to which are given in the bibliography at the end of the book.

As in the preceding chapters on signal analysis, no attempt at mathematical rigour has been made. Some analysis has been included to clarify the principles involved, but the interested reader is again urged to study one of the standard works on this subject. The intention in this chapter is to provide sufficient understanding of correlation analysis to enable the reader to make intelligent use of a laboratory correlation analyser.

SIGNAL CLASSIFICATION

Signals may be classified as two types: deterministic and random. When a signal can be described explicitly in terms of mathematical relationships it is deterministic. An example of a deterministic signal is the displacement of a one degree of freedom vibrating system, which can be calculated at any instant once the initial conditions and dynamic characteristics of the system are known.

More generally, signals are likely to be random. In other words, we cannot predict their value at any instant in time. A record of wind-induced structural vibration, for example, is likely to have random characteristics.

Deterministic signals can appear to be random if they are corrupted by the presence of noise or distortion. Signals which appear random can often be made to yield deterministic data after suitable analysis, such as autocorrelation. Alternatively, it is often found that unwanted periodic components (such as mains hum) are included in a random signal. The periodic components can be removed by suitable analysis such as digital filtering.

Deterministic signals either have a periodic form or are transients. Periodic signals can be analysed by the use of Fourier series, and transients by the use of the Fourier Transform. As we saw earlier, the spectra resulting from periodic signals consist of discrete lines, while the spectra obtained from transients are continuous (since in theory there are an infinite number of frequency components in a transient).

Probabilistic and statistical techniques must be used to describe a random signal. The general procedure is to use a set of N separate time histories (usually called records). The set of records is known as an ensemble. The mean value can be obtained at an instant t_1:

$$\bar{x} = \frac{1}{N} \sum_{i=1}^{N} x_i(t_1) \tag{9.1}$$

The simplest way to describe a random signal is to calculate its mean square value from equation (9.2):

$$\bar{x}^2 = \lim_{T \to \infty} \frac{1}{T} \int_0^T x^2(t) \, dt \tag{9.2}$$

Alternatively the root mean square (RMS) value can be used. Both of these give a description of the signal's amplitude, but no information which may help to explain the variable nature of the process. To do this the probability of the signal amplitude falling between certain bounds must be determined.

The probability P_1 that the signal amplitude will lie within the range x to $x + \delta x$ is normalised by δx to give the probability density function:

$$P_1(x) = \lim_{T \to \infty} t_n / (T \delta x) \tag{9.3}$$

(t_n is the fraction of the total record time T that the signal lies between x and $x + \delta x$). As the record length is extended to infinity, equation (9.3) becomes the probability distribution function:

$$P(x) = \int_{-\infty}^{\infty} P_1(x) \, dx \tag{9.4}$$

As we shall see in the next section, the mean value of the product of samples taken at two separate times t_1 and t_2 for each record in the ensemble is called the autocorrelation. A random signal is said to be stationary if the mean value x and the autocorrelation $R_{xx}(\tau)$ (where $\tau = t_2 - t_1$) are constant for all values of t_1 and t_2. One way to test whether a random signal is stationary or not is to take a long record and partition it into a number of sections of equal length. If the mean value and the autocorrelation function obtained from each partitioned section are the same as those calculated from the complete record, the signal is said to be an ergodic random process. All the statistical properties of such a process may be obtained by analysis of a single record.

AUTOCORRELATION

We have seen that spectral analysis techniques such as the FFT lead to a description of a waveform in terms of its energy content at different frequencies. Correlation can best be thought of as a measure of the similarity between two waveforms, or between a waveform and a delayed version of itself. The correlation between two waveforms can be evaluated by multiplying the two waveforms together at every instant in time, and summing all the products. If the two waveforms are identical as in figure 9.1(a) each product will be positive, and the resulting sum will be large. On the other hand, if we have a pair of waveforms which are dissimilar some of the products will be positive and some negative as shown in figure 9.1(b). The cancellation that results when the sum is evaluated means that the final correlation will be smaller than in the case of the identical waveforms.

If we take a waveform and a time-shifted version of itself as shown in figures 9.2(a) and (b), we can define an autocorrelation function for the waveform. If the time shift between the two records is zero we have the condition shown in figure 9.1(a)—that is, the two records are identical and the final sum of the products (the correlation) will be large. If the time shift is not zero the two records appear dissimilar and the correlation is small.

We can extend the analysis further by finding the average product for each time shift, by dividing each correlation sum by the number of contributing products. If this average product is plotted as a function of the time shift the result is known as an autocorrelation function. The autocorrelation of a random signal is largest when the time shift is zero, and diminishes as the time shift increases.

The above discussion gives an explanation of autocorrelation in words, but as good engineers we shall of course want to see some mathematical justification. We have defined the autocorrelation function as the mean value of the product of two samples taken at times t_1 and t_2, where the delay τ between the record and a delayed version of itself is

$$\tau = t_1 - t_2$$

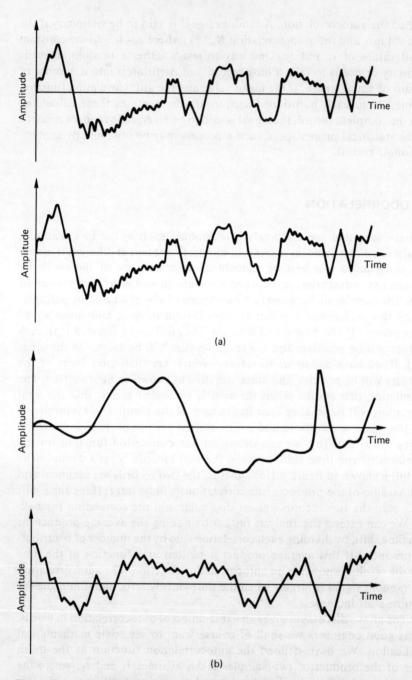

Figure 9.1 (a) Multiplying two identical functions together at every point and summing the products results in a large correlation value. (b) Multiplexing dissimilar waveforms together at every point and summing the products results in a low correlation value.

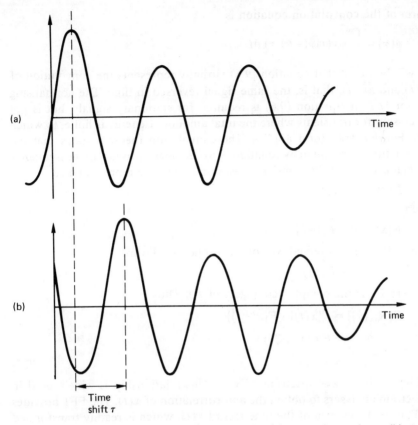

Figure 9.2 Correlation of time-displaced signals: (a) original waveform; (b) time-shifted waveform.

The autocorrelation function is therefore

$$R_{xx}(\tau) = \frac{1}{N} \sum_{i=1}^{N} x_i(t_1) x_i(t_2) \tag{9.5}$$

If we take an infinite number of products from a stationary ergodic process, that is, let $N \to \infty$, the autocorrelation function becomes

$$R_{xx}(\tau) = \lim_{T \to \infty} \frac{1}{T} \int_0^T x(t_1) x(t_2) \, dt$$

(where the record length is T) (9.6)

Since we can never have $T = \infty$, $R_{xx}(\tau)$ is estimated by

$$\frac{1}{T} \int_0^T x(t_1) x(t_2) \, dt$$

The convolution theorem states that a Fourier Transform (either forward or reverse) transforms a convolution into a product and vice versa. One form of the convolution equation is

$$g(\tau) = \int_{-\infty}^{\infty} x(t) y(-t+\tau)\, dt$$

It will be noted that equation (9.6) virtually represents the convolution of $x(t)$ and $x(-t)$, that is, the same signal reversed in time. The normalising factor $1/T$ in equation (9.6) is required for stationary signals, but is not necessary for transients where the total amount of energy is finite, to which the Fourier Transform applies. The convolution theorem states that the correlation represented by equation (9.6) becomes a product in the frequency domain. If we let the Fourier Transform of $x(t)$ be represented as

$$F\{x(t)\} = F_x\{f\}$$

then

$$F\{x(-t)\} = F_x\{-f\}$$

(since if time is reversed, vectors will rotate backwards)

$$= F_x^*(f)$$

(where F^* is the complex conjugate of F). Thus

$$\begin{aligned}F\{R_{xx}(\tau)\} &= F\{x(t)\} \cdot F\{x(-t)\} \\ &= F_x(f) \cdot F_x^*(f) \\ &= |F_x(f)|^2 \end{aligned} \qquad (9.7)$$

which is the power spectrum of $x(t)$. This relationship is widely used in spectrum analysers to obtain the autocorrelation of $x(t)$. The FFT provides a Fourier Transform of the time record $x(t)$, which is readily transformed into a power spectrum by multiplying each frequency domain point by its complex conjugate. A Reverse Fourier Transform is then used to get the autocorrelation function.

The most important properties of the autocorrelation function are listed below:

(a) $R_{xx}(0) = \bar{x}^2$, that is, the autocorrelation for $\tau = 0$ is the mean square value of the signal.
(b) $R_{xx}(\tau) = R_{xx}(-\tau)$, that is, autocorrelation is an even function. This is because the correlation of $x(t)$ and $x(t+\tau)$ is the same as the correlation of $x(t)$ and $x(t-\tau)$. For this reason it is conventional practice to display only the positive half of the function.
(c) $R_{xx}(0) \geq R_{xx}(\tau)$, or in other words the autocorrelation function is a maximum at $\tau = 0$.

INTERPRETING AUTOCORRELATION DIAGRAMS

The main value of the autocorrelation function is that it can reveal periodicity in a signal which otherwise appears to be random. However, before

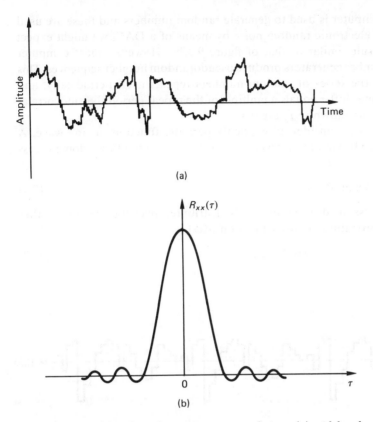

Figure 9.3 Wideband random noise autocorrelation: (a) wideband random noise; (b) autocorrelation function (theoretical) for wideband random noise.

considering the autocorrelation of a periodic waveform let us look at the wideband random noise signal shown in figure 9.3(a). Let the spectrum of the random noise be uniform over a wide bandwidth B, that is

$$F_x(f) = G \quad \text{for } 0 \leq f \leq B \text{ (where } G \text{ is spectral density)}$$
$$= 0 \quad \text{otherwise}$$

The autocorrelation function can then be shown to be

$$R_{xx}(\tau) = GB\left\{\frac{\sin(2\pi B\tau)}{2\pi B\tau}\right\} \tag{9.8}$$

This has the same form as the familiar function $\sin(x)/x$, and we see that the autocorrelation function diminishes rapidly as we move away from $\tau = 0$ and has its first crossing point at $\tau = 1/(2B)$. The autocorrelation function (shown in figure 9.3(b)) has a maximum at $\tau = 0$ as we should expect, but note that even a slight increase in τ is sufficient to reduce the magnitude considerably.

If a computer is used to generate random numbers and these are used to produce electronic random noise by means of a DAC, we might expect to see a result similar to that of figure 9.3(b). However, most computer random-number generators produce pseudorandom number sequences. This means that the series of random numbers repeats itself periodically, and when the time shift τ equals a multiple of the repeat period the autocorrelation repeats as shown in figure 9.4.

Let us now consider an explicitly periodic function; a sine wave. A sine wave can be viewed as the result of a stationary ergodic random process with sample records:

$$x(t) = A \sin(\omega t + \phi) \tag{9.9}$$

where ϕ is assumed to be uniformly distributed over 0 to 2π, or in other words the probability density function $p(\phi)$ is

$$p(\phi) = 1/2\pi \qquad (0 \leq \phi \leq 2\pi) \tag{9.10}$$

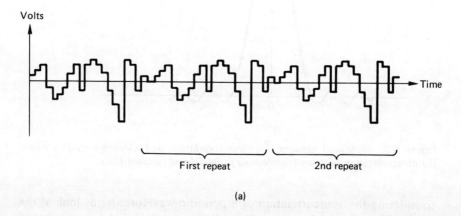

(a)

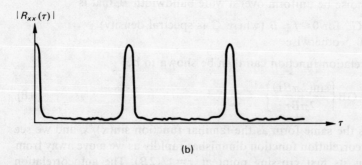

(b)

Figure 9.4 (a) Computer-generated random noise sequence (repeated 3 times). (b) Autocorrelation of pseudorandom noise.

CORRELATION AND SPECTRAL ANALYSIS 211

From equation (9.6) the autocorrelation function of a sine wave is given by

$$R_{xx}(\tau) = \lim_{T \to \infty} \frac{1}{T} \int_0^T x(t)x(t+\tau) \, dt \tag{9.11}$$

Substituting from equation (9.9) and noting that the averaging is over ϕ as given by equation (9.10), it follows that

$$R_{xx}(\tau) = \frac{A^2}{2} \int_0^{2\pi} \sin(\omega t + \phi) \sin(\omega(t+\tau) + \phi) \, d\phi$$

$$= (A^2/2) \cos(\omega \tau) \tag{9.12}$$

Hence the autocorrelation function of a sine wave is a cosine function with

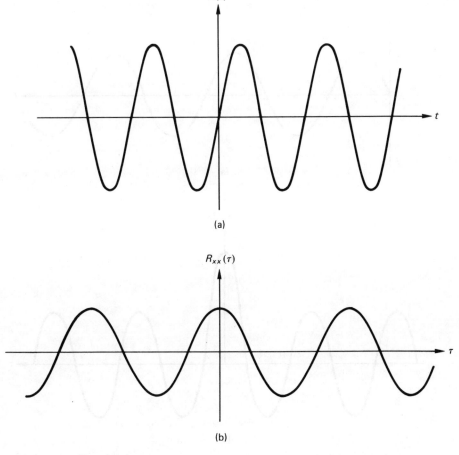

Figure 9.5 Autocorrelation of periodic function (sin x) is also periodic: (a) x(t) = sin(x): (b) autocorrelation of sin(x).

212 INSTRUMENTATION FOR ENGINEERS

an amplitude equal to the mean square value of the original sine wave, as illustrated by figure 9.5. The key observation to be made from figure 9.5(b) is that the envelope of the autocorrelation function remains constant over all time delays. This tells us that future values of the data may be predicted precisely from past observations. If we now look back at the autocorrelation function for wideband random noise shown in figure 9.3(b), we see that the envelope of the correlation function diminishes rapidly. This tells us that knowledge of the exact time history of the random process shown in figure 9.3(a) will not help us to predict future values.

Finally, let us consider the case of a signal consisting of a mixture of a sine wave and some wideband random noise. The waveform is

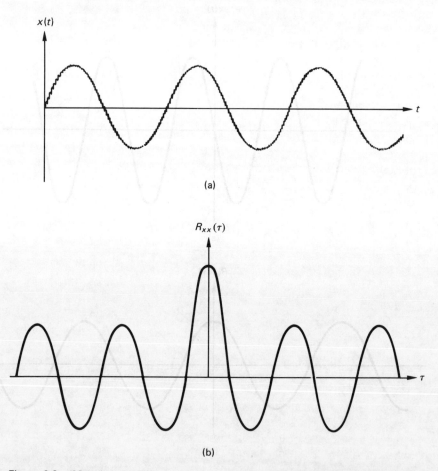

Figure 9.6 'Noisy' sinusoid and its autocorrelation: (a) sine wave plus random noise; (b) autocorrelation of (a).

shown in figure 9.6(a). The autocorrelation function in this case is simply the sum of the autocorrelations of the sinusoid and the wideband noise, as given by equations (9.12) and (9.8). That is

$$R_{xx}(\tau) = \frac{A^2}{2}\cos(\omega\tau) + GB\left(\frac{\sin(2\pi B\tau)}{2\pi B\tau}\right) \tag{9.13}$$

where A is the amplitude of the sine wave and G is the spectral density of the random noise. Figure 9.6(b) shows a typical correlation function of this type. The autocorrelation decays quickly from a maximum value at $\tau = 0$ to the cosine term describing the sinusoidal portion of the data. An autocorrelation function of this type suggests that it is possible to predict values into the distant future with limited precision.

CROSS-CORRELATION

The autocorrelation function is a measure of the similarity between a signal and a time-shifted version of itself, and as we have seen it can be used to detect periodicity in a signal. An obvious extension of the technique is to use correlation to measure the similarity between two non-identical waveforms. This is called the cross-correlation function. If the same signal is present in both waveforms it will be reinforced in the cross-correlation function, while any random noise will be reduced through cancellation. Cross-correlation is commonly used in radar and sonar, where the transmitted waveform is cross-correlated with the received signal to recover the 'echo'.

For a pair of stationary ergodic processes the cross-correlation function can be defined as a time average:

$$R_{xy}(\tau) = \lim_{T\to\infty} \frac{1}{T} \int_0^T x(t)y(t+\tau)\,dt \tag{9.14}$$

By analogy with the autocorrelation function, $R_{xy}(\tau)$ is a measure of how two signals co-vary at different points on the time axis. However, unlike the autocorrelation function $R_{xy}(\tau)$ does not usually have its maximum value at $\tau = 0$. For example, if $x(t)$ is a sonar pulse and $y(t)$ a delayed echo, the cross-correlation function $R_{xy}(\tau)$ might look like figure 9.7 in which the time τ_{max} corresponds to the mean time delay relating the two signals.

The cross-correlation function may be obtained via the Fourier Transform as was the case for autocorrelation. The procedure is first to calculate a cross-power spectrum from the two time domain signals $x(t)$ and $y(t)$. A cross-power spectrum is defined as the product of the individual Fourier

214 INSTRUMENTATION FOR ENGINEERS

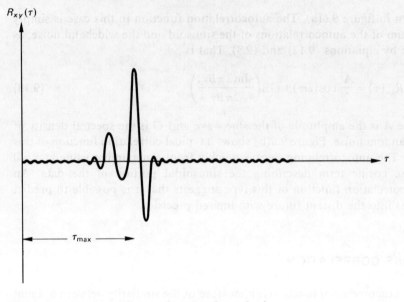

Figure 9.7 Cross-correlation of sonar pulse and echo.

Transforms $F_x(f)$ and $F_y(f)$, that is

$$F_{xy}(f) = F_x^* \cdot F_y(f) \tag{9.15}$$

where $F_x^*(f)$ denotes the complex conjugate of $F_x(f)$. The cross-correlation function $R_{xy}(\tau)$ is obtained by applying a Reverse Fourier Transform to $F_{xy}(f)$, that is

$$R_{xy}(\tau) = F^{-1}[F_{xy}(f)] \tag{9.16}$$

The most important properties of the cross-correlation function are:

(a) $R_{xy}(\tau) = R_{yx}(\tau)$.
(b) If $R_{xy}(\tau) = 0$ for all τ, the signals $x(t)$ and $y(t)$ are said to be uncorrelated. This usually means that they arise from two completely unrelated processes.
(c) $|R_{xy}(\tau)|^2 \leq R_{xx}(0) R_{yy}(0)$. This inequality is often used to monitor the maximum possible value of the cross-correlation. The normalised cross-correlation defined by

$$r_{xy}(\tau) = \frac{R_{xy}(\tau)}{[R_{xx}(0) R_{yy}(0)]^{1/2}}$$

is often used instead of $R_{xy}(\tau)$ since it scales the cross-correlation in terms of the root mean square values of the signals $x(t)$ and $y(t)$.

INTERPRETING CROSS-CORRELATION FUNCTIONS

The most straightforward use of cross-correlation is in analysing propagation problems, where it is required to determine to what extent a signal measured at a point originates from a particular source, and with what time delay. Of particular interest are multiple path problems, where $x(t)$ can propagate through several possible paths to give an output signal $y(t)$ as shown in figure 9.8. Assuming that the propagation is wideband and non-dispersive

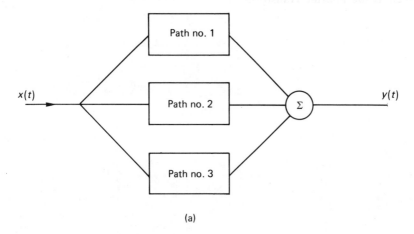

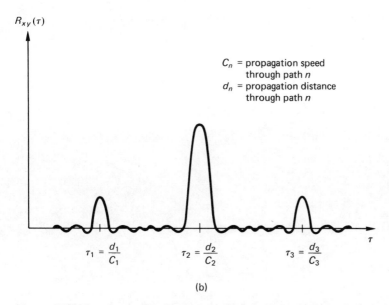

Figure 9.8 *Use of cross-correlation in multiple path propagation analysis: (a) non-dispersive signal propagation through multiple paths; (b) cross-correlation.*

216 INSTRUMENTATION FOR ENGINEERS

(that is, the velocity does not depend on frequency), a cross-correlation measurement will in theory yield a collection of peaks as in the example of figure 9.8(b). Each peak defines the contribution through a given path. In practice, however, it is often found that some of the smaller peaks may be obscured by noise.

Cross-correlation can also be used to assess the contribution made by a number of signal sources to a single output signal. However, in general, problems of this type are better handled by spectral analysis techniques such as the Fourier Transform.

Bibliography and Further Reading

CHAPTER 1

J. Topping, *Errors of Observation and their Treatment*, 4th edn, Chapman and Hall, 1972.
R. K. Penny, *The Experimental Method*, Longman, 1974.
P. R. Bevington, *Data Reduction and Error Analysis for the Physical Sciences*, McGraw-Hill, 1969.
Open University Science Foundation Course Team, *The Handling of Experimental Data*, Open University Press, 1970.

CHAPTER 2

B. E. Jones, *Instrumentation, Measurement and Feedback*, McGraw-Hill, 1978.
M. J. Usher, *Sensors and Transducers*, Macmillan, 1985.
L. F. Adams, *Engineering Measurements and Instrumentation*, English Universities Press, 1975.
B. A. Gregory, *An Introduction to Electrical Instrumentation and Measurement Systems*, Macmillan, 1973.

CHAPTER 3

Strain Measurements, Brüel and Kjaer.
J. M. Calvert and M. A. H. McCausland, *Electronics*, Wiley, 1978.
H. Ahmed and P. J. Spreadbury, *Analogue and Digital Electronics for Engineers*, 2nd edn, Cambridge University Press, 1984.
P. Horowitz and W. Hill, *The Art of Electronics*, Cambridge University Press, 1980.
Z. H. Meiksin and P. C. Thackeray, *Electronic Design with Off-the-shelf Integrated Circuits*, Parker Publishing Company Inc, 1980.

CHAPTER 4

M. E. Van Valkenburg, *Analogue Filter Design*, Holt, Rinehart and Winston, 1982.
R. A. Williams, *Communications Systems Analysis and Design*, Prentice-Hall, 1987.

CHAPTER 5

As chapter 3.

CHAPTER 6

P. Horowitz and W. Hill, *The Art of Electronics*, Cambridge University Press, 1980.
A. Colin, *Programming for Microprocessors*, Butterworth, 1979.
D. Aspinall, *The Microprocessor and its Application*, Cambridge University Press, 1978.

CHAPTER 7

R. D. Stuart, *An Introduction to Fourier Analysis*, Methuen, 1966.
K. G. Beauchamp and C. K. Yuen, *Digital Methods for Signal Analysis*, Allen and Unwin, 1979.
J. S. Bendat and A. G. Piersol, *Random Data: Analysis and Measurement Procedures*, Wiley, 1971.

CHAPTER 8

R. B. Randall, *Frequency Analysis*, Brüel and Kjaer, 1977.

CHAPTER 9

C. Chatfield, *The Analysis of Time Series*, 2nd edn, Chapman and Hall, 1980.
K. G. Beauchamp, *Signal Processing*, Allen and Unwin, 1973.
J. S. Bendat and A. G. Piersol, *Measurement and Analysis of Random Data*, Wiley, 1966.
J. S. Bendat, *Principles and Applications of Random Noise Theory*, Wiley, 1958.
J. S. Bendat and A. G. Piersol, *Engineering Applications of Correlation and Spectral Analysis*, Wiley, 1980.

Index

54 series 120
74 series 120

AC amplifier 64–5
AC bridge 42, 53–5
AC tachometer 23–4
Accelerometer 13, 24–5, 41
Acoustic sensors 38–9
Active filter 79–101
Actuator 2
A/D/A converters 103
ADC 102–17
Address decoders 133, 145
Aliasing 96, 102, 186–90
Analogue filters 79–101
Analogue multiplexers 115–16
Analogue spectrum analyser 180–2
Analogue-to-digital converter *see* ADC
Antialiasing 102, 186–90
Attenuator 64
Autocorrelation 203–13

Bandpass filter 5, 79–80, 85–6, 89–90, 96–7
Bandstop filter 80–1
Bandwidth 56, 60
Baud 157
BCD 138
Bearing 161
Bidirectional buffers 132–3
Binary arithmetic 119, 135–8
Binary code 135–8
Binary coded decimal 138

Bit 136
Boolean algebra 119, 121–5
Breaking strain 27, 28
Bridge circuit 29, 42–55
Bus transceivers 132–3
Butterworth filter class 82–3
Byte 136

Cable capacitance 72
Campbell diagram 167–8
Cantilever vibration 164–7
Capacitive displacement transducers 16–18
Capacitive sensors 16–18
Capacitor microphones 38
Carbon-track potentiometer 16
CARRY flag 136
Central processing unit 140–2
Centronics port 118
Ceramic microphones 38
Charge amplifier 72
Chebyshev filter class 82–3
Choice of sampling parameters 187
Circle diagram 137
CMOS circuits 121
CMOS outputs 119–20
Combinational logic 121–5
Combining random and systematic errors 10–12
Common Mode Gain (CMG) 58–9
Common Mode Rejection Ratio (CMRR) 59
Comparator (analogue) 73

INDEX

Comparator (digital) 133–4
Comparison of window functions 197–8, 201
Compensation (strain gauges) 27, 29, 45–50
Complement, of a number 136
Complex form of the Fourier series 177
Compliance 71
Constant current source 65
Convolution 207–8
Cooley and Tukey's FFT algorithm 185
Corner frequency 80, 95–6
Correlation 205–16
CPU 138–9, 140–2
Cross-correlation 213–16
Current-to-voltage converter 71

D/A converter see DAC
DAC 102–3, 105–9
DC bridge 42
DC tachometer 23
Deterministic signal 203–4
DFT 184
Differential amplifier 65–6
Differentiator 70–1
Digital device families 118–21
Digital filter 79
Digital spectrum analysers 182
Digital-to-analogue converter see DAC
Digitisation 102
Dirac delta function 164
Dirichlet's conditions 170
Discrete Fourier Transform see DFT
Displacement sensing 13–21
Distortion 61, 161
Dual slope integration 112–13
Dummy gauge 29, 45–6
Dynamic range 96, 202

Electret microphones 38
Electrical resistance strain gauge 25–7
Ensemble 204
Equal-ripple filter see Chebyshev filter class
Ergodic process 205
Error analysis 6–12
Even and odd functions 173–4

Fast Fourier Transform (FFT) 184–8
Fatigue life (of strain gauges) 27–8
Filter class 82–3

Filter design 84–97
Filter order 81–2
First-order bandpass filter 85–6
First-order filters 84–6
First-order high-pass filter 85
First-order low-pass filter 84–5
Flash encoder see Parallel encoder
Flow sensors 29–32
Forward transform 178
Fourier analysis 170
Fourier cosine series 173
Fourier series 170–7
Fourier sine series 174
Fourier transform 177–9
Fourier's integrals 172
Fourier's theorem 170
Fourth-order high-pass filter 94
Fourth-order low-pass filter 93
Frequency domain analysis 159
Frequency-to-voltage converters 109
Full bridge 49–52

Gain–bandwidth product 60, 61
Gain–frequency diagram 60
Gauge factor 26–8, 45
Gauge resistance 26–8
Gaussian function 7, 12
Gaussian window 198, 201
Generalised instrumentation design 3–6
Goodness-of-fit tests 12
Gray code 18

Half-bridge 46–9
Hall effect sensor 39–40
Hamming window 197–8, 201
Handling precautions (CMOS) 121
Handshaking 157
Hanning window 197–8, 201
Heisenberg's Uncertainty Principle 4–5
Hexadecimal notation 138
High-pass filter 79–80
Hot-wire anemometer 30
HPIB see IEEE Interface
Hydrophone 39

IEEE Interface 118, 157–8
Impedance bridge 53–5
Impulse see Dirac delta function
Inductive sensors 53–5
Input bias current 59–60

INDEX 221

Input impedance 58
Input–output impedance 58
Integrating ADC 112–13
Integrator 68, 70
Interrupt service routine 144
Inverting amplifier 62–3

Linear moving magnet transducer 22–3
Linear Variable Differential Transformer 14–15
Lissajous figure 68
Load cell 28–9
Logic identities 123, 125
Longitudinal velocity sensing 21–3
Low-frequency turbulence 39
Low-pass filter 79–80
LVDT see Linear Variable Differential Transformer

Magnetic microphone 38
Magnetic velocity sensing 21–4
Mass flow rate sensor 30
Maximally flat filter see Butterworth filter class
Mean 7, 204
Mean square value 204
Mechanical strain gauges 25
Memory 139–40
Microphones 38–9
Microprocessor 138–44
Microprocessor interfacing 118–58
Microprocessor timing diagram 150–1
Modal domain 164–7
Modal response diagram 167
Modifier 5
Modulating transducers 5–6
Moving coil velocity sensor 21–2
Multiple path problems 215–16
Multiplexed ADC 115–16

Natural binary 135–6
Noise immunity 119
Non-inverting amplifier 63–4
Normalised filter design 84
Number codes 135–8
Nyquist frequency 186, 199

Op-amp see Operational amplifier
Op-amp rules 62
Operational amplifier 55–73
Optical displacement sensors 18–21

Optical gratings 21
Optical sensors 35
Output impedance 58

Parallel analysers (real time) 181–2
Parallel encoder 110–12
Parallel plate capacitor 17–18
Parallel-to-serial conversion 134–5
Perfect filters 80
Peripheral interface 142–3
Photovoltaic cells 36–7
Piezoelectric effect 24
Piezoelectric sensors 24–5, 38
Pistonphone 39
Pitot tube 30–1
Poisson's ratio 49, 52
Potentiometer displacement transducers 15–16
Power spectrum 208, 213
Precision rectifier 67–9
Pressure sensors 28–30
Primary sensing element 3
Probability density 204
Pseudorandom noise 210
Pulse counter 144–51
Push–pull sensor 53–4

Quarter-bridge 44–6

Random access memory (RAM) 139
Random errors 7–9
Random noise 204–13
Read-only memory (ROM) 139
Real filter transfer functions 80
Recovery time 67, 73
Resistance-wire thermometer 33
Resistive bridges 42–52
Resonance 24–5, 164
Reverse Fourier Transform 178
RF5609A 95–6, 100–1
RM5604 96–9
RM5605 96–9
RM5606 96–9
Root mean square 204
Rotary LVDT 15
Rotary potentiometer 16
Rotational velocity sensing 23–4
RS232 157
RS432 157

Sample-and-hold devices 109–10
Sampling 102

Saturation 70, 73
Scratch gauges 25
Second-order bandpass filter 89–90
Second-order filters 86–90
Second-order high-pass filter 88–9
Second-order low-pass filter 86–8
Self-exciting sensors *see* Self-generating sensors
Self-generating sensors 5
Semiconductor strain gauges 27–8
Semiconductor temperature sensors 33–4
Serial-to-parallel conversion 134–5
Shaft whirl 168
Shannon's sampling theorem 188–90
Signal classification 203–5
Signal conditioning 2, 55
Signal conversion 102–17
Signed arithmetic 136
Significance tests 12
Simple bending, measurement of strain in 46–7, 50–1
Slew rate 60–1
Slip rings 23
Special-purpose filter devices 95–7
Spectrum analysers 180–8
Spring-mass system 159
Standard deviation 7
Standard error 7
State parameters 2
Stationary signals 205
Stepped filter analysers 180
Strain gauge 25–9
Strain gauge transducers 28–9
Stress 47–9
Successive approximation ADC 113–15
Summing amplifier 66–7
Swept-filter analyser 181
Switched-capacitor filter 95
Systematic error 9–10

Tachometer 23–4
Temperature sensors 32–4
Tensile strain measurement 47–9, 51–2
Thermistor 33–4
Third-order filter 90
Third-order high-pass filter 92–3
Third-order low-pass filter 90–2
Time domain 159
Timing (microprocessor) 150–1
Tracking converter 115
Transient response of filters 82
Tri-state buffers 131
TTL circuits 119–20
TTL inputs 119–20
TTL outputs 119–20
Turbine flow sensor 31–2
Two's complement 138

UART 134–5
Unidirectional buffer 131
Unity-gain voltage follower 63

Variable permittivity capacitive transducer 17
Variable-area capacitive sensor 17–18
Variable-separation capacitive sensor 17–18
Vector flow 30–1
Vector flow transducer 30–1
Vector response diagrams 168–70
Virtual earth 63
Voltage Supply Rejection Ratio (VSRR) 59
Volume flow sensors 31–2
VRD *see* Vector response diagrams

Waterfall diagram 167–8
Wheatstone bridge 42–3
Windowing 190–9
Windshields 39
Wire and foil strain gauges 26
Wire-wound potentiometer 16
Word 136

X register 142

Zero error 6